WORLD BANK TECHNICAL PAPER NUM

Toward Participatory Research

Deepa Narayan

The World Bank
Washington, D.C.

Copyright © 1996
The International Bank for Reconstruction
and Development/THE WORLD BANK
1818 H Street, N.W.
Washington, D.C. 20433, U.S.A.

All rights reserved
Manufactured in the United States of America
First printing April 1996

Technical Papers are published to communicate the results of the Bank's work to the development community with the least possible delay. The typescript of this paper therefore has not been prepared in accordance with the procedures appropriate to formal printed texts, and the World Bank accepts no responsibility for errors. Some sources cited in this paper may be informal documents that are not readily available.

The findings, interpretations, and conclusions expressed in this paper are entirely those of the author(s) and should not be attributed in any manner to the World Bank, to its affiliated organizations, or to members of its Board of Executive Directors or the countries they represent. The World Bank does not guarantee the accuracy of the data included in this publication and accepts no responsibility whatsoever for any consequence of their use. The boundaries, colors, denominations, and other information shown on any map in this volume do not imply on the part of the World Bank Group any judgment on the legal status of any territory or the endorsement or acceptance of such boundaries.

The material in this publication is copyrighted. Requests for permission to reproduce portions of it should be sent to the Office of the Publisher at the address shown in the copyright notice above. The World Bank encourages dissemination of its work and will normally give permission promptly and, when the reproduction is for noncommercial purposes, without asking a fee. Permission to copy portions for classroom use is granted through the Copyright Clearance Center, Inc., Suite 910, 222 Rosewood Drive, Danvers, Massachusetts 01923, U.S.A.

The complete backlist of publications from the World Bank is shown in the annual *Index of Publications*, which contains an alphabetical title list (with full ordering information) and indexes of subjects, authors, and countries and regions. The latest edition is available free of charge from the Distribution Unit, Office of the Publisher, The World Bank, 1818 H Street, N.W., Washington, D.C. 20433, U.S.A., or from Publications, The World Bank, 66, avenue d'Iéna, 75116 Paris, France.

ISSN: 1014-9848
ISBN: 0-8213-3473-5

Deepa Narayan is a social scientist in the Social Policy and Resettlement Division of the World Bank's Environment Department.

Library of Congress Cataloging-in-Publication Data

Narayan-Parker, Deepa.
    Toward participatory research / Deepa Narayan.
        p.   cm. — (World Bank technical paper; ISSN 0253-7494; no. 307)
    Includes bibliographical references
    ISBN 0-8213-3473-5
    1. Water resources development—Research—Developing countries—Citizen participation. 2. Community development—Research—Developing countries. I. Title. II. Series.
HD1702.N372   1995
333.91'13'091724—dc20   95-45755
                                                                         CIP

Contents

Foreword v

Abstract vi

Preface vii

Acknowledgments viii

Section I 1

1. Introduction 3

2. What Is Participatory Research? 17

3. Defining the Purpose of the Study 33

4. Organizing for the Participatory Research Cycle 47

5. Choosing Data Collection Methods 59

6. From the Field: Innovative Data Collection Methods 81

7. Short Cuts to Sampling 101

8. Selection and Training of Field Workers 113

9. Data Analysis, Dissemination and Use 129

Section II 139

10. Participatory Research Activities 141

Section III 227

11. Participatory Research Checklists 229

Annexes 255

1. Terms of Reference: Process of Identification of a Rural Development Project by Its Future Beneficiaries 257

2. Bibliography and Selected References 263

Foreword

Development strategies directed at poverty alleviation are undergoing a fundamental transformation. With each passing year, a gathering momentum is shifting the traditional, centrally managed, supply-driven emphasis of development toward demand-based approaches that place local stakeholders at the heart of the process. In the past, communities were expected to make a contribution to a government or agency undertaking. Today, increasingly, governments and external support agencies are expected to provide support for community-centered initiatives.

Information gathering and data collection are central to ensuring that external assistance meets the needs and priorities of poor urban and rural communities. When participatory research methods are utilized, genuine ownership of both the process and the outcome results. By involving users in project development and implementation, participatory research helps build local capacity to solve problems and make sound decisions. This, in turn, leads to an improved chance that facilities will be used and maintained on a sustained basis.

Participatory research is a dynamic, demand-based and change-oriented data collection methodology which helps to address the new challenges facing development practitioners. Using participatory approaches, trained professionals can work with community members to gain insight into the local social and cultural context, the goals and perspectives of various actors, and most important, the "felt needs" of affected stakeholders. While participatory research places new demands on the researcher, it also increases the likelihood that the results of the data collection will be effectively applied to bring about needed change.

A commitment to poverty alleviation requires creative responses on the part of the World Bank and other development agencies. In this regard, this volume, the third in a series on participatory development produced by the PROWWESS/UNDP-World Bank Water and Sanitation Program, provides many insights, innovative tools and techniques. Based on hard-won field experience gained largely in the water and sanitation sector, these methods are proving to be applicable in other sectors and settings as well.

A growing body of evidence suggests that the kinds of tools and approaches advocated in this volume and its companions can make a sizable contribution to putting poor people in charge of improving their health, their environment and their living standards. These tools and approaches should be used widely by development practitioners in both national and international agencies. We hope this volume and its companions will contribute to this goal.

Ismail Serageldin
Vice President
Environmentally Sustainable Development

Abstract

If people in communities are to take greater initiative and responsibility for their own development, they must be actively involved in decisionmaking. In order to successfully implement an approach emphasizing user involvement, planners must understand the social fabric in which change will be embedded. Data collection for social analysis can provide planners information crucial for achieving desired objectives.

Participatory research is a process of collaborative problem solving through the generation and use of knowledge. It is dynamic, demand-based, and change-oriented, and seeks to raise people's awareness and capacity by equipping them with new skills to analyze and solve problems. When carried out in a participatory manner involving users in decisionmaking, the data collection process itself can foster ownership and involvement in follow-up actions. This, in turn, can lead to more effective, fully used, and better maintained water and sanitation systems.

Participatory research embodies an approach to data collection that is two-directional — both *from* the researcher to the subject and *from* the subject to the researcher. In conducting participatory research, the role of managers shifts from executing blueprints and plans to institutionalizing a problem solving capacity within agencies and communities. Effective managers not only create an environment in which information flows quickly within an agency, but also from agencies to communities and from communities to agencies.

This document is intended to be used by technical and social science staff involved in planning and implementation of water and sanitation facilities for the poor. It may also be useful to social science researchers applying their craft to the low cost water supply and sanitation sector. With adaptation, the methods have application across sectors.

In addition to identifying the principles underlying participatory research — and how these differ from conventional research methodologies — the document provides guidance on defining the purpose of a study, organizing the participatory research process, choosing from available data collection methods, selecting and training field worker to carry out the research, dissemination and use of findings. Drawing upon experience gained during the past fifteen years in more than twenty countries, the volume also offers a wide array of participatory research activities and checklists.

Preface

This volume, *Toward Participatory Research*, is designed as a practical guide to the formulation and implementation of participatory research and inquiry. It contains a wealth of material about the principles underlying participatory techniques, insights gained from the use of such techniques in the field (noted as boxes complementing the text), suggestions about how best to design and implement these methods, and actual participatory activities and checklists which have proven successful.

Section I provides an in-depth examination of the essential issues that confront anyone interested in or committed to participatory research. These include:

- What participatory research is and how it differs from more conventional data gathering;

- The advantages and disadvantages of participatory techniques;

- The role of the participatory researcher in the process, and how to select and train field workers to adopt a participatory stance;

- How to define and limit the purpose, scope and objectives of a participatory study;

- Which techniques to employ in different circumstances; and

- How to analyze and disseminate findings to ensure use by decisionmakers.

Section II contains detailed information on thirty-three participatory activities — community profiles, gender analysis, semi-projective techniques, games and simulations, technology-related activities, and management and problem identification tools — which have been developed and used around the world. These activities are spelled out sufficiently so that they can be used by practitioners in community and agency settings.

Section III is made up of a number of useful checklists the researcher can employ to ascertain information ranging from community economic and social factors to how services are managed by an agency.

Those readers who have already employed participatory processes in their work may choose to spend relatively less time on Section I and explore in greater depth the activities found in Section II. Conversely, those new to the subject are urged to review thoroughly the basic premises underlying participatory research before attempting to use it in a field setting. In this regard, the boxes scattered throughout the first nine chapters marked "Field Insight" may prove particularly useful.

Acknowledgments

This document has gone through a seven-year evolutionary process, and has benefitted from the experience and support of numerous agencies and individuals. The earliest draft was written during a three-month stay at the International Reference Center (IRC), Netherlands in 1987. The document was expanded while I was with the PROWWESS program at the United Nations Development Programme in New York in 1989. The final evolution occurred over the last two years, through a collaborative effort between the United Nations Development Programme-World Bank Water and Sanitation Program, and the Social Policy and Resettlement Division of the Environment Department of the World Bank. Publication was made possible by financing from the government of Norway.

I would particularly like to acknowledge Hans Van Damme and Jan Teun Visscher (IRC) and Siri Melchior-Tellier and Lyra Srinivasan (PROWWESS) for championing participatory approaches to research long before these approaches were supported by international organizations.

During the past two years, John Blaxall, David Kinley and Bruce Gross of the UNDP-World Bank Water and Sanitation Program have been particulary helpful in ensuring the completion of a series of publications on participatory development, of which this is the final volume. The present document owes much to the organization and editorial skills of Ted Howard, supported by Laurie Edwards, Mary Mahy, Lyra Srinivasan and Ellen Tynan. The document draws heavily on my personal field experience, the experience of many nongovernmental organizations (NGOs), my social scientist colleagues within the Bank, and the Program field staff, particularly Ron Sawyer and Mary Judd. All boxes within the text not otherwise attributed are based on the author's experiences.

This volume has benefited from comments received from many people, in particular: Han Heijnen, IRC; Phil Evans, ODA; and Sandy Davis, Mary Judd, Wendy Wakeman, and Barbara Parker, of the World Bank.

Photo Credits: Asem Ansari 4 (top), 18 (top); Curt Carnemark 10, 21, 41, 57, 72, 112, 228; Laurie Edwards 15; IRC 56 (bottom); Sam Joseph 52 and 53 (all); David Kinley 11, 118; Annie Manou 68, 82 (bottom), 153; Deepa Narayan 2, 4 (bottom), 8, 9 (all), 13, 14, 16, 20 (both), 22, 31, 32, 35, 40, 46, 58, 60, 65, 71 (bottom), 75 (top), 80, 82 (top and middle), 83, 86 (both), 87 (both), 90, 104 (bottom), 105, 111, 123, 136, 140, 148, 184 (middle and bottom), 197, 200 (both), 207; Dan Owen 91, 102; Ron Parker 208; Jake Pfohl 19, 75, 76, 78, 96, 99, 104 (top); Robert Pini 149 (both), 150; Ron Sawyer 5, 18 (bottom), 23, 28, 38, 44, 71 (top), 95, 122, 197; Betty Soto 62, 85, 184 (top); Lyra Srinivasan 61; Chris Srini Vasan 69; UN Photo Library 185. All other photographs courtesy of the PROWWESS Program.

SECTION ONE

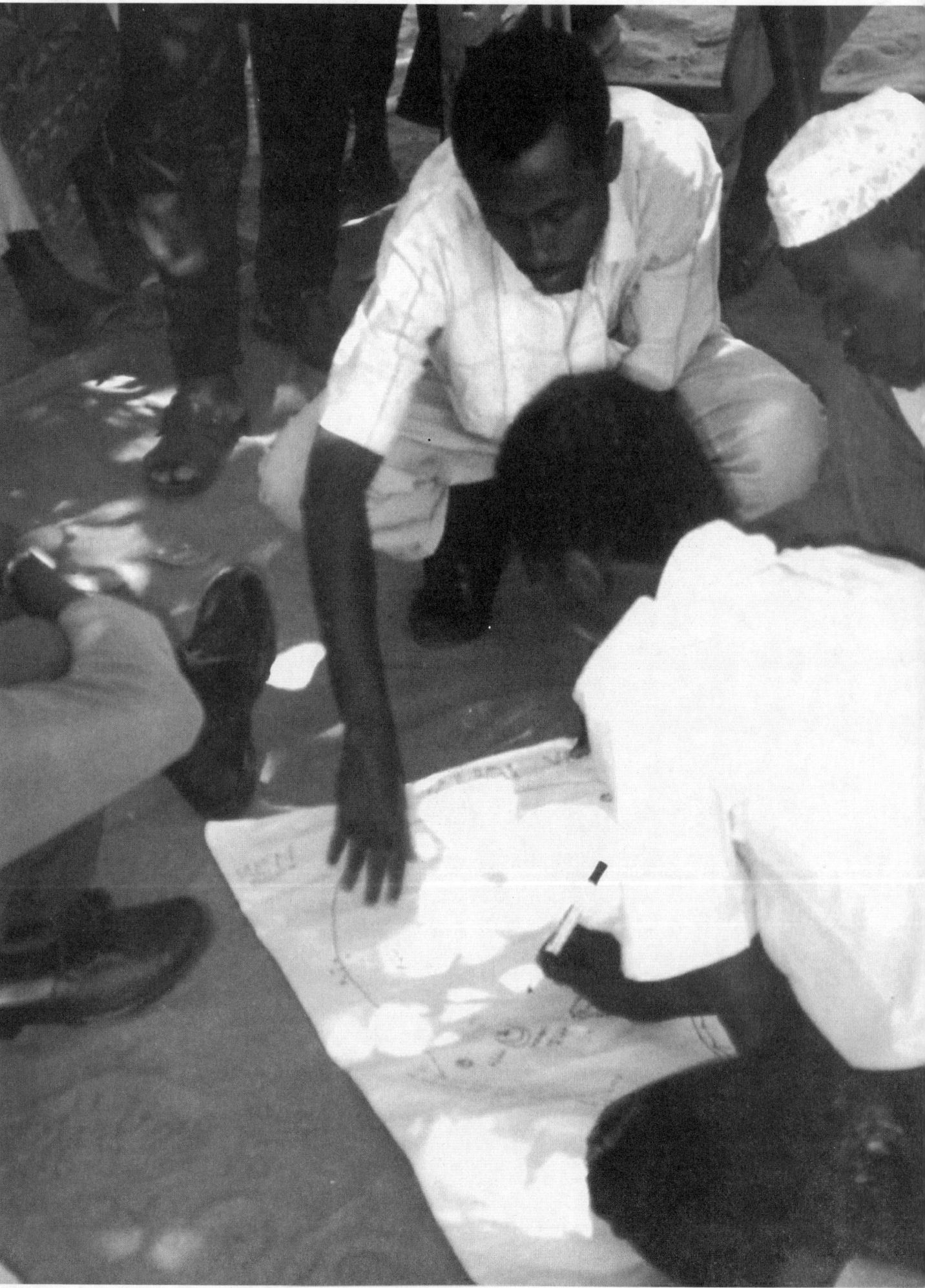

CHAPTER 1

INTRODUCTION

All managers need data to perform effectively, assemble resources, develop appropriate plans, and implement them properly. Managers of community-based water supply and sanitation programs are no different.

The last decade marked a change in the management of these programs, as emphasis shifted from centrally-managed, supply-driven approaches to demand-driven, community partnership approaches. With this shift in emphasis from community assistance in government initiatives to government assistance in community initiatives, agencies must now provide an enabling environment for thousands of different communities to manage, sustain, and effectively use safe water and sanitation facilities. This shift has profound implications for program and project staff. Planning and implementing this changed role is the new challenge faced by agency managers of water and sanitation programs and projects.

If people in communities are to take initiative, be creative, learn, and assume responsibility for their own development, they must be actively involved in decisionmaking. Involved users, in turn, are more likely to commit their time, resources, and abilities toward selecting, paying for, installing, using, and maintaining water supply and sanitation facilities in the long run.

In order to successfully implement an approach emphasizing user involvement, planners can no longer focus narrowly on technological factors. Managers must understand the social fabric in which the technology will be embedded: the human dimension affecting community acceptance, use, and how the management of technologies is organized. This is particularly important when the goal is to reach the poor.

Thus, previously "irrelevant" factors — social organization, networks, community groupings; local leadership and rivalries; who controls access to different resources; who fetches water and how it is fetched; where and why people prefer to defecate or use certain types of sanitation facilities — become very relevant. Technical assessment must be complemented with social analysis to achieve sustainability. In short, planners must understand the individual and group factors operating at the community level which affect the functioning, use, and organization of water and sanitation technologies.

The process of data collection for social analysis can provide planners with crucial information about human and organizational factors so that the chosen technology fits the existing social environ-

Chapter Contents

This chapter explores the following issues:

- Why participatory processes in data collection are important
- Key issues in the community-based approach to the water and sanitation sector
- Purpose of this document
- Principles of participatory research and inquiry
  - Capacity building
  - Utilization of results
  - Short-cut methods
  - Multiple methods
  - Expertise of the nonexpert

ment and achieves desired objectives. When carried out in a participatory manner involving users in decisionmaking, the data collection process itself can foster ownership and involvement in follow-up actions. At the community level, this, in turn, can lead to more effective, fully used, and better maintained water and sanitation systems.

The role of agencies in supporting community initiatives is quite different from the traditional role of engineering agencies. Hence, institutional analysis is also needed to identify the capacity of agencies to interact in a partnership mode with communities. Based on this, appropriate institutional arrangements can be developed.

Planning for a community-based water and sanitation system is an iterative process, one which decentralizes decisionmaking away from the agency to the community and the field office. This requires frequent adjustment based on lessons learned in the field. As information flows from the community upward, changes are made to support options and develop strategies that help in achieving the overriding goal of sustainability and effective use of water and sanitation systems.

Community decisionmaking means that programs do not come into existence with blueprints, but evolve with field-based learning. In such situations, information flow becomes more important than when implementing pre-established plans. A different type of information is needed, as well.

Data collection for programs based on a community partnership approach is important for:

- Assessing felt needs or demand for improved services;

- Understanding the social and cultural context, social organization and conflict resolution mechanisms of groups and communities;

- Clarifying goals and objectives of different actors;

- Identifying gender-based differences in access to resources, and in developing strategies to empower women;

- Developing management and institutional strategies to support community and user decisionmaking;

- Conducting institutional analysis of community organizations, intermediaries, and national support agencies;

- Developing process and outcome indicators to monitor and evaluate local decisionmaking and capacity development.

It must be emphasized, however, that no matter how participatory the approach to data collection and research, there is still a critical role for the trained professional. Even arriving at the place

where the community can become involved requires professional support, guidance, and judgment. The most critical decision to be made is defining the purpose of the study and limiting the possible universe of data to be collected; in this, expert guidance can be essential.

Key issues in the community-based approach to the water and sanitation sector

In water and sanitation projects, the data collection process must be guided by the key issues relevant to the sector. While the data collected will vary according to the circumstances and to the purposes of the study, two broad categories of issues must be considered in all situations.

The first category — *matching community demand with agency supply* — focuses on identifying community demand and assessing the client orientation of support agencies. At the community level, felt need is measured by willingness to pay for different service levels, to commit other non-cash resources, and to organize for collective action as needed. At the agency level, capacity to undertake a flexible approach supportive of community needs and capacities must be assessed, and suitable institutional arrangements subsequently made. This may require the involvement of NGOs and the private sector, or government departments familiar with community organizational work.

The second category — *defining sustainable management systems* — involves examining a range of technologies based on organizational, social, cultural, ecological and financial factors. The long-term sustainability of the technology system depends on how operation and maintenance is organized at the community level and the linkage between the community and the outside world. Hence, local management capacity to sustain the technology system is of critical importance.

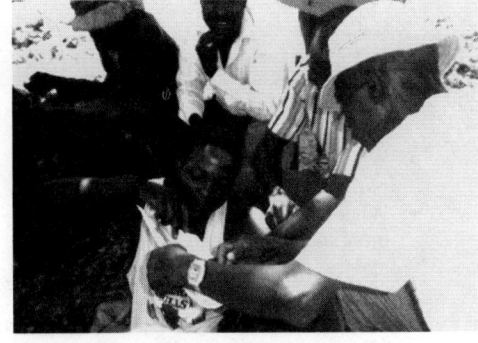

The structure and function of most water and sanitation utilities developed at a time when projects were mainly seen as straightforward "design and build" undertakings. Engineers worked in environments that were highly predictable and controllable. They depended on blueprints and plans, and on centralized, hierarchical structures. They were evaluated based on their adherence to construction schedules and their achievement of physical targets.

The conventional blueprint method remains appropriate and effective in many circumstances, but does not fit situations which require partnership with hundreds of communities. Partnership implies equal status, mutual respect, two-way information exchange, negotiation, and shared decisionmaking.

To work in close cooperation with large number of communities, government agencies have to build flexibility into their organization.

Data Collection and Project Planning

Project planning is an iterative and cyclical process with two very distinct categories of issues that must guide the data collection process.

Category 1: Matching community demand with agency supply

Criteria for selecting groups, communities and agencies:
- What is the willingness to pay among users?
- Are community groups willing to commit resources and able to organize themselves?
- Are the poor targeted?
- What is the capacity of the agency to respond to local demand?

Category 2: Designing sustainable management systems

Criteria for determining technology options and local management capacity:
- What service levels are desired by users?
- What is the local management capacity?
- What is the capacity of the agency to provide a range of service levels?
- What is the availability of spare parts and technicians?
- What support will be provided to build local management capacity, and by whom?

They have to be able to adapt and change to fit local culture, indigenous knowledge systems, organizations and local needs.

Since no two communities are alike, joint decisionmaking with communities means working in environments that cannot be standardized and are relatively unpredictable and uncontrollable. Yet, no manager in a government agency can operate only in response to community initiatives and in reaction to total unpredictability.

Hence, the role of managers in community-based projects is to manage unpredictability. This means reducing the unknown and unpredictable to manageable proportions without imposing inappropriate structures prematurely. This can be done by designing a learning environment. The primary role of managers thus shifts from executing blueprints and plans to institutionalizing a problem solving capacity within agencies and communities. This ensures that programs can evolve with changing circumstances.

Experience from around the world across sectors (irrigation, health, agriculture, to name a few) shows that managing a "learning environment" is heavily dependent on a two-way flow of information, systematic data collection, analysis, and use to guide decisionmaking. Effective managers not only create an environment in which information flows quickly within an agency, but also from agencies to communities and from communities to agencies. Experience shows that when engineers accept and become proponents of a participatory, community-based approach, fundamental reorientation of agencies is possible.

Indeed, projects that do not come into being with a blueprint are in danger of going off-track unless those responsible for them have a high degree of self-awareness and a commitment to adjustment. This can be guided by monitoring of the most important processes and benefits throughout the project cycle.

To function effectively and efficiently, managers of community-based projects need different types of contextual information than do managers of conventional water and sewerage projects. A timely and relevant flow of information will help managers tolerate higher levels of ambiguity while resisting the temptation to impose too much structure prematurely.

Purpose of this document

There are many ways of collecting data. The purpose of this document is to provide guidance on participatory research and inquiry. Emphasis is also placed on short-cut methods for applied research so as to provide decisionmakers at the community and agency levels with timely information.

The material contained in this volume represents a simplified application of social science research for use by development practitioners. It is a mix of simple, pragmatic and participatory techniques that can be applied at any step of a policy or project cycle.

While each study is unique, the underlying processes and issues that must be resolved are common to all studies. Hence, this document centers on the processes involved in developing, conducting and utilizing the methods of participatory data collection. Rather than promoting any particular technique, a problem solving approach is utilized to focus on the issues, options and implications of options at each stage of a study. Guidance and suggestions are provided through examples distilled from experiences in the field.

This document is intended to be used by technical and social science staff involved in planning and implementation of water and sanitation facilities for the poor. It may also be useful to social science researchers applying their craft to the low cost water supply

Field Insight: Stakeholders Identify Institutional Requirements of a Community Management Approach, Indonesia

The WISPLIC project for water and sanitation for low income communities, designed with assistance from the World Bank, is the first Bank-financed national project using a demand-driven project strategy and a learning process approach in Indonesia. In effect, it consists of six provincial projects, all devolving authority to lower levels of government and to the community.

As part of the pre-appraisal process, a two day workshop was held in one of the outer islands which brought together sixty senior government personnel from the provinces and the center with consultants and World Bank staff. The workshop was facilitated by a local consultant, the head of an NGO and a Bank staff member.

Brief presentations were intermixed with hands-on activities. Most work was done in small groups. The first activity consisted of participants drawing their personal visions of community management on large sheets of paper. These were then presented and discussed in the plenary session. Following the presentations, participants next examined the roles and responsibilities that would be required to fulfill each vision. Cards were distributed listing different decisions that needed to be made in any water project, and five levels of decisionmakers — the central government, provincial, district, sub-district and community. Participants discussed what decisions would be made at which levels to support their vision.

What emerged from the long, often heated discussion was agreement that the simpler the technology being used — such as spring captures and improved wells — the greater the agreement among the participants to give control and authority to community groups. This was followed by an activity focused on the behavioral attributes (such as leadership and initiative), skills and training needed to enable communities and government staff to take on new roles of responsibility for selection and approval of projects. Subsequently, 15 to 20 percent of project costs were earmarked for capacity building. The common vision emerging from the workshop proved very useful in evaluating preliminary proposals submitted by consultants and to judge the approaches being tested in "pilot villages."

and sanitation sector. With adaptation, the methods have application across sectors.

Companion volumes to this work are also available. A complementary document on participatory evaluation and a tool kit of activities to foster participatory development have been produced by the PROWWESS group of the UNDP-World Bank Water and Sanitation Program. As a set of inter-related materials, they are designed to assist in carrying out training, planning, evaluation, and research to support local decisionmaking.

Principles of participatory research

Participatory research and inquiry is based on certain principles influencing how studies are conducted and how their results are used. The essential underlying principles are:

- Capacity building
- Utilization of results
- Short-cut methods
- Multiple methods
- The expertise of the nonexpert

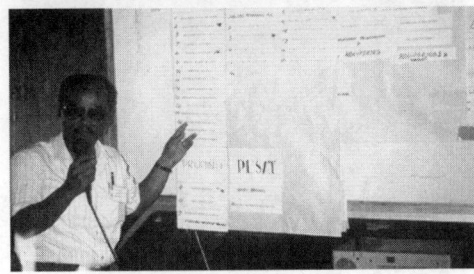

Capacity building

A key principle of participatory research is the importance of empowerment — strengthening local capacity to conduct problem analysis on its own in the future. Hence, the focus is on user involvement in the data collection process.

Utilization of results

In participatory data collection, the main goal of the research process is to ensure that the data will be used appropriately. A study that is methodologically sound and has collected much relevant information is useless if the findings are not utilized. Accordingly, the study process at each stage includes user involvement to increase interest and commitment to the study and its results. User involvement ranges from planners in provinces or country capitals, to men, women and children in communities.

Short-cut methods

Most studies operate under resource and time constraints. It is more useful to have rough information made available to planners when they need it rather than having information accurate to the "nth" degree available six months later when the programs have already been developed, and the questions that are being asked have changed. It is therefore often important to use short-cut methods that yield reliable and relevant information.

Multiple methods

It is important to include different perspectives and different methods in studying a problem, especially when short cuts are being taken. This helps ensure that the information is complete and reliable.

The expertise of the non-expert

The abilities, intelligence, and knowledge of local staff and community members are relied upon in the research process. People — including children — are usually knowledgeable about their environment; their interest, abilities, preferences, and knowledge should be respected and utilized.

The analogy of the photographer

Participatory research is the process of systematic problem solving. A participatory research study is an integrated process through which information needs are defined, analyzed, and reported, and

Field Insight: Community Learning and Research, Kenya

The rural water sector in Kenya is a paradox: while the value of community participation is conventional wisdom, successes are still limited. In order to learn from various community-based approaches while giving priority to community learning through participatory research, the Regional Water and Sanitation Group in Nairobi developed a two-pronged strategy to distill lessons and build local capacity.

The first strategy focused on facilitating community learning and growth through communities conducting their own "research." RWSG staff contacted communities managing their own pipe or point water systems in different parts of Kenya and arranged for exchange visits and discussions between communities. During these five-day exchanges, community groups questioned project management, visited community user groups including women's groups, and visited shops or people that stored spare parts. Much of the discussion focused on the organization needed to keep water systems functioning. Questions asked were challenging, and mutual advice was shared on water sales, leadership wrangles, stubborn non-paying members and need for supportive income generating activities. As a result of the exchanges, communities which were not charging for water began considering water charges. The additional immediate effect was that after the visiting groups saw the success of the income generation activities in Kakamega, they formulated plans to initiate such activities in their own communities.

In the second strategy — documentation — researchers developed a set of analytical case studies distilling lessons from the same programs based on detailed discussions with communities and site visits.

Source: *Apollo Njonjo and Patrick Obora Okoth, personal communication, RWSG, Nairobi, 1993.*

---

follow-up action is taken, to fulfill clearly defined purposes. Such a study is an organic whole with many parts, each influencing the next and being influenced by that which preceded it. Formulating and conducting such a study requires making multiple decisions at various levels, all of which influence each other.

The planner or manager of a participatory study is analogous to a photographer who is on assignment to shoot a photo essay of a community. With a limited supply of film, the photographer must carefully plan what situations to focus on, what lens to use, where to stand relative to the subject, and on what aspect of the community to concentrate in order to tell the story. Planning the shoot in advance, the photographer must be ready to adapt, adjusting the shooting strategy to accommodate or to take advantage of the unexpected and unplanned.

No single picture will give a composite understanding of the community being studied. Standing rooted in one place, the photographer's perspective becomes limited by the unchanging nature of her position. When taking pictures of people, rather than objects or landscape, the photographer is presented with new challenges, because people may react to her presence in unpredictable ways. Accordingly, she must learn to adopt strategies to minimize the negative effects created by her own involvement.

Even the process of developing and printing the pictures presents a new set of issues and dimensions. By employing chemical processes in different ways, and by the use of different darkroom techniques, the photographer can highlight certain features, bringing them to the forefront of attention.

So it is with the researcher. Like the photographer with a limited supply of film, participatory researchers do not have the luxury of unlimited time and resources. By carefully defining the purpose of the study, they know what issues are important, what must be focused on, and what types of information are vital.

Researchers must decide whether they will collect a range of information with little depth or detail, or data within a relatively narrow range but with more depth. They must also decide on acceptable levels of accuracy: is information on the number and type of water sources sufficient, or is data required on history, operation, and maintenance of each source; is it enough to conclude that most adult women do not use the community toilet or is it necessary to state precisely that 65.8 percent of the women do not use the facility?

Depending on the type, desired accuracy, and depth of information, the researcher makes important decisions about how to collect data, who should collect it and from whom. The relative sensitivity of the information will also impact the process.

Key Questions for Program Analysis and Project Design

The primary challenge in designing large-scale programs is supporting local capacity and achieving a fit between community felt need or demand and agency response. Thus the primary issues to be considered in analysis or design of new programs are:

Does the program or project design respond to community demand?
Does the process support local capacity development?

Data collection that assists in suggesting strategies or evaluating the following eight questions is particularly important.

1. Sub-project or community selection criteria- Are communities selected based on demand? Is a community self-selection process in place?

2. Demand assessment- How is demand assessed and how is it integrated in the sub-project selection process and implementation?

3. Institutional framework- What are the institutional arrangements, and what is the capacity to implement a community based approach?

4. Funds and financial flow- Are the rules and regulations governing the release and flow of funds, such that they will respond in a timely way to community demand?

5. Technology and service levels- Do people have a choice of technologies and service levels?

6. Planning: Master-plan or a learning process approach- Is the planning based on a learning process approach with built-in flexibility?

7. Women's empowerment- Is empowerment of women through their involvement an overall goal?

8. Monitoring and evaluation- Do the key indicators of success reflect local capacity development and is a system of monitoring process and outputs in place and used for decisionmaking?

Adapted from: *Deepa Narayan, 1994. The Contribution of People's Participation: Evidence from 121 Rural Water Supply Projects. ESD Occasional Paper Series, 1. World Bank.*

Participatory Research is an Organic Whole

Purpose of study

- What is the purpose of the study?
- Who will do the study?
- What are the resource constraints, personnel, facilities, money?
- What is the timeframe?

- What information will be collected - why?

- Who will collect information - Why?
- Who will give information - Why?
- How will information be collected - Why?
- When will information be collected - Why?
- Where will information be collected - Why?
- How long will information collecting take - Why?
- How much will information collection cost - Why?

- Who will analyze information - Why?
- How will information be analyzed - Why?
- When will information be analyzed - Why?
- Where will information be analyzed - Why?
- How long will analysis take - Why?
- How much will it cost - Why?

- Who will present results - Why?
- To whom will results be presented - Why?
- How will results be presented - Why?
- When will results be presented - Why?
- Where will results be presented - Why?
- How long will it take to prepare results - Why?
- How much will it cost - Why?

Results

- Who are the primary users of results - Why?
- What results or type of information do they need - Why?
- What are the other implications of results - Why?
- When are results needed - Why?
- How will results be utilized - Why?
- Where will results be utilized - Why?

The processes used to analyze data must be appropriate to the nature of the data collected, keeping in mind the skills and tools available and the needs and interests of the potential users of the results. Conversational interviews can be analyzed by theme; background information can be described and summarized in a table; decisionmaking within local institutions can be described.

This then leads to decisions about how to present results to ensure that they will be utilized by different categories of users. Data can be presented through a short descriptive narrative report, a report using tables and diagrams, or panel discussions, workshops and seminars or visuals and role plays.

For the photographer, a good "how to" book can assist in picture taking by showing through diagrams and words how a camera works, what types of lenses, film, tripods, flashes, filters and cameras are available, and how and when they should be used. But in the end, the aspiring photographer learns by doing — by choosing equipment, selecting shots, and deciding on how to print the photograph. For this reason, a good photographer is one who not only knows the mechanics of photography, but is creative in applying skills.

So it is with participatory research. A book, such as the present volume, can describe the tools and their potential uses. But a successful study depends on the creative application of these tools in culturally and professionally appropriate ways that fulfill the specific purposes for which the research was undertaken. This is the great challenge and opportunity each of us must face.

Chapter 2

What is Participatory Research?

There are two broad approaches to research — *conventional research* and *participatory research*.

Conventional research is characterized by "experts" — outsiders or those external to a situation — who gather quantitative and qualitative information about people, a community, an agency or a situation without the research subjects being involved in the process. The approach could be described as a one-way flow of information from the subject to the researcher. The process is relatively static, one in which information is gathered from a community or agency, and then processed and analyzed by experts for their use with little or no feedback to the community or agency.

In contrast, *participatory research* embodies an approach to data collection that is two-directional (both from the researcher to the subject, and from subject to researcher). The process itself is dynamic, demand-based and change-oriented.

The participatory approach identifies and involves all those persons, agencies and organizations with a substantial stake in an issue. This includes women, men and children in communities, especially from marginalized groups, but also agency staff, policymakers, and all those affected by decisions made through the participatory research process. A central goal of the process is to involve people as active creators of information and knowledge. This is done not only because it results in the inclusion of different and important interests, needs and perspectives, but because it also increases the chances of the findings being put to use.

Participatory research seeks to raise people's awareness and capacity by equipping them with new skills to analyze and solve problems. This is achieved by involving people in the development of every step of the research process, rather than by having them follow predetermined research methods imposed from the outside. As a result, the distinction between the roles of the external researchers and the "subjects" — the people being studied — should become less pronounced. External experts and professional interact with community members or a project agency primarily as facilitators.

By way of illustration, contrast how community mapping would likely be carried out through these two distinct approaches.

Depending on the number of the communities to be studied, a traditional researcher might approach the mapping process by:

Chapter Contents

This chapter explores the following issues:

- How conventional and participatory research/data collection differ
- Characteristics of participatory research
  - Process
  - Collaboration
  - Problem solving
  - Knowledge generation
- The methodology of participatory research
  - Credibility
  - Trustworthiness
  - Relevance
  - Feasibility
- Advantages and disadvantages of participatory research
- Roles and characteristics of a participatory researcher
  - Creative technician
  - Human research instrument
  - Facilitator
  - Trainer/enabler
  - Communicator
  - Advocate or activist

- Satellite imagery;

- Aerial photography (relatively low-cost methods have been developed);

- Land surveys using professional cartographers; or,

- Mapmaking conducted by field workers who walk through a community and seek assistance from key local people.

While each of these approaches will likely result in an accurate map of the area, none involves the community. As a result, local people are not engaged in learning further about their own community, nor do they have the chance to contribute their own knowledge and experience to the study.

By contrast, a participatory approach to mapping communities might utilize one of the following approaches:

- A mapping activity played with school children who are asked to begin by drawing their own home and those of their immediate neighbors;

- In working with communities, a facilitator could ask a group of men and women to use locally available materials — clay, grass, mud, stones and sticks — to make a model of their ward or community and mark all water sources;

- During a group meeting, a facilitator might initiate discussion and involve people in the process of developing a map of their community using paper or crayons;

- Satellite maps can be taken to the community for interpretation and used for planning multi-community activities such as watershed management.

These participatory approaches serve three different purposes:

- By being more involved in the process of mapping, people become more aware of their environment;

- Researchers learn about people's "cognitive maps" and what features of the community are considered most important by the people who live there; and

- Information is elicited that can be used for further planning both by the participants and the external facilitators.

All research is concerned with the reliability and validity of information. In conventional research, guidelines have been developed and documented for use with each specific technique. These rules and guidelines are available on how to get reliable and valid information through questionnaires, interviews, naturalistic observa-

---

Conventional and Participatory Approaches to Data Collection

When to use the conventional approach

1) When data needed are mostly quantitative.
2) When follow up action, in terms of program and project training, is uncertain.
3) When issues addressed are not sensitive.
4) When the purpose of the study does not include setting the stage for staff or community involvement in a program.
5) When time and resources are not serious constraints.

When to use the participatory approach

1) To establish rapport and a commitment to use study results.
2) When staff or community interest and involvement is central to achieving program goals.
3) When information is complex or sensitive.
4) When major issues are unknown or relatively undefined.
5) When supporting local capacity is important.

tion, focused observation, discussion groups, and to a lesser extent through games and simulations.

By its very nature, participatory research does not operate by clear-cut rules handed down to data collectors by experts. Rather, guidelines for data collection are arrived at with study participants through consensus, reflection, dialogue and experience. Validity of information is best established by involving users at all levels — from villagers to managers — in interpreting the information.

Clearly, there is a fundamental difference between the conventional approach to data collection and the participatory approach to research. Which approach one uses depends on the purpose of the study to be carried out. For example, participatory approaches at the community level may be inappropriate for project identification studies. Involving the community in research at this early stage might raise expectations of assistance where none may be forthcoming, at least for several years. Similarly, a conventional approach might be inappropriate when information on sensitive issues is needed quickly, community participation is central to program viability, staff cooperation is essential, or when issues relevant to a particular cultural setting remain unclear. (For a summary of the differences between conventional and participatory research, see page 30.)

Field Insight: Developing Toilet Designs Through Models, the Maldives

A research team including a ministry engineer and a health educator conducted an evaluation of the national public toilet program in the Maldives combined qualitative, quantitative and participatory techniques. Much information was collected through focus group discussions and site visits. In addition, a pre-determined interview format with open-ended questions was used to interview users and non-users about what they liked and disliked about using toilets. This information was subject to content analysis and quantified and tabulated.

Two wooden models with movable parts were used with groups of men and women. One model represented the existing community latrine which consisted of a block of four latrines in a row. Four cubicles built around a shallow well was the second model. People could move the model's doors, roof, and windows, add or subtract new features. Discussion focused on design of the community toilets. People expressed their preferences by playing with the wood models adding and subtracting features, until they were satisfied. By the end of fifteen group discussions, clear patterns emerged for toilet design. These were incorporated into the new national program for sanitation.

Characteristics of participatory research

Participatory research is a process of collaborative problem solving through the generation and use of knowledge. It is a process that builds local capacity by involving users in decisionmaking for follow-up actions.

The end objective of participatory research is pragmatic: to solve problems. Hence its methods and techniques are not bound by the protocols and conventions of particular academic disciplines.

The single most important criterion guiding the process is ensuring utilization of the research. Vital methodological issues are centered on who controls research and how utilization of its findings can be assured. This has profound implications for how research is conceptualized and how the research process of data collection, analysis, dissemination, and planning for follow-up action is conducted.

The approach and methods of participatory research have evolved out of experiences in the field. They are a result of liberal borrowing from different disciplines, including psychology, anthropology, sociology, adult education, economics, statistics and philosophy of science. Principles and techniques from these disciplines have been distilled, tested, and refined based on what has worked.

Research in this pragmatic context is a creative rather than a mechanistic process in which learning never ceases. The process is characterized by flexibility, responsiveness, adaptation and inventiveness. While the methodology of this approach is not unique, what distinguishes it is the application of existing methods and the creation of new techniques in ways that encourage participation and involvement.

It must be clearly recognized that participatory research is not a pure art form or an "all or none" phenomenon. Although the ideal is shared decisionmaking, in reality this varies on a continuum from high to low levels of participation. Indeed, different degrees of participation may be desirable in different political and cultural contexts or at different stages in a program cycle.

Depending on the nature and timing of the study during the life cycle of a program, it may be possible to combine elements of a participatory approach with those of a traditional approach. Whatever degree of participation may be achieved in a particular context, participatory research has certain characteristics that distinguish it from other research. These characteristics include a focus on: process, collaboration, problem solving, and knowledge generation.

Process

The process of undertaking participatory research is more important than the output per se or the methods used. This process is characterized by collaboration between different levels of users, participants or interest groups, those participating in a program or those influenced by the decisions made in a program. It includes project "beneficiaries" as well as program and project staff and funders. Special efforts are made to ensure that those traditionally overlooked — women, children, the poor, and junior project personnel — have opportunities to become centrally involved in the research process if they choose to do so.

Because no one should be forced to collaborate or participate, people are often self-selected or self-defined by continuing to choose to participate in the research decisionmaking process. However, this self selection is valid only if everyone has genuinely been encouraged through appropriate mechanisms in a supportive, non-threatening environment. For example, because the timing is inconvenient, village meetings held in the early morning rarely attract women. Similarly, in most contexts, extension workers will be unlikely to speak up in the first meeting with senior management staff or policymakers.

Any research process has to make decisions about data collection methods. In the participatory research process, the key is how the decisions are made, rather than what decisions are made or what specific methods are used. No particular method is by definition excluded or considered better than others. The adequacy of a method depends on the context. However, there is usually a preference for short-cut methods to conventional approaches.

Collaboration

Participatory research involves shared decisionmaking, which also implies shared control and power among participants, clients and potential users. Without noticeable degrees of collaboration with users during the research phases, research cannot be considered participatory. Asking people to fill out questionnaires or to participate in answering questions posed by an outside interviewer does not qualify as participatory research. Involvement of decisionmakers in the research process increases ownership of the results, their credibility, and the probability of their use.

Without the collaboration of local people in communities, researchers have no access to indigenous knowledge, learning systems, or different perspectives on the realities of community life. At the agency level, project managers and staff become the substantive experts. More than the researcher, they know about the context within which the research will be undertaken and utilized. This includes the context of resources, institutions, administration, culture, and politics.

Problem solving

As noted above, the purpose of participatory research is to contribute toward understanding of a problem, thereby leading to action and problem resolution. Participatory research is not an end in itself but

---

Field Insight: Participatory Research in Emergency Rehabilitation, Maharashtra, India

The experience of the Government of Maharashtra with support from the World Bank demonstrates the power of participatory data collection and planning in housing and village construction following devastation by an earthquake. Initially layout designs for new villages were developed by urban architects and planners who were determined to build villages "better than most towns". This involved a grid layout. A two-week planning workshop involving architects, planners, administrators and village people, facilitated by an experienced participatory trainer, changed the planning approach and village designs. The work included site visits and discussion groups with villagers, men, women and youth.

Although people initially wanted the grid design (which was the only design they had seen in the newly rebuilt villages), discussion revealed dissatisfaction with the grid. Older people complained that the new village outlays disrupted the social groupings based on religion and caste as well as destroyed the privacy of the houses. In the old villages the majority of the houses did not open on the streets whereas in the new grid lay out the entrance to most houses opened directly on the streets. The village layouts were redesigned by architects and local people working together.

Source: *Meera Shah, World Bank, 1994*

a means to problem solving. This focus on problem solving ensures that research yields data that are relevant to decisionmakers and hence have a greater likelihood of being utilized.

Knowledge generation

Generating information in the age of computers is easy, but it is difficult to transform information into knowledge that is relevant to people's actual situation and needs.

When community members and agency staff are involved in the research process — from identifying the issues to be studied to interpreting and disseminating results — the process of engagement results in learning. This learning, in turn, can lead to changes in people's "cognitive maps", and consequently to new ways of understanding a situation and how to effectively act to improve upon it.

The methodology of participatory research

All researchers are concerned with the validity and reliability of their findings. For the participatory researcher, whose approaches are eclectic and may involve creating or experimenting with new techniques, methodological rigor has a new meaning. *Reliability* is achieved by using multiple methods and *validity* confirmed through consensus, discussion and dialogue.

In the participatory research approach, there can be no absolute definition of a "perfect" study or "best" methods. Rather, what the researcher is focused on is the issue of "appropriateness," identifying and using the appropriate methodology for a specific situation and context. The best participatory studies are those that are utilized because they address the needs of the users in a particular context and yield reliable and valid information.

Field Insight: Stakeholder Involvement, Planners, Engineers and Statisticians

Participatory research methods with their focus on the "expertise of the nonexpert" and innovation in data collection tools, are often resisted by government agency planners, engineers and statisticians. Sometimes, this is because of lack of familiarity with participatory approaches and their potential. However, once exposed to participatory methods in action, the same planners, engineers and statisticians become powerful allies and proponents of the participatory research process.

In Lesotho, the manager of a national sanitation program became a convert when he accompanied a team to the districts to test participatory evaluation tools to reduce the load on centralized and computerized monitoring systems. For example, instead of a socio-economic household survey, the manager observed, and then led, a game of "open-ended snakes and ladders" to evaluate children's knowledge of health and sanitation principles after completion of a health education program. (See Section II for details of the open-ended snakes and ladders game.)

In Kenya, discussions with senior economists, planners and statisticians of the Ministry of Planning and Central Bureau of Statistics not only piqued their interest but resulted in their joining the research team in the field for different lengths of time. Subsequently, some of the statisticians became so involved that the Bureau requested additional training for staff in qualitative and visual tools. (See Section II for detailed information on such tools as gender analysis, health seeking behavior and wealth ranking.)

In Pakistan, the attendance of senior engineers, together with social scientists in a participatory training workshop, resulted in a changed orientation of several rural water programs.

In India, MYRADA, an NGO, now conducts training for Indian Administrative Services (IAS) officers by taking them to villages for up to a week of hands-on training aimed particularly at changing attitudes and behavior. The work includes trying out tools such as mapping, transects, and matrix ranking. While IAS officers will probably never conduct such activities themselves in villages, the experience makes them proponents of planning with local people based on shared power and control in future programs.

The most important criteria for designing a participatory study and selecting the methods to be employed are credibility, trustworthiness, relevance and feasibility.

Credibility

Both the study design and the methods used must have credibility in the eyes of stakeholders and decisionmakers; results which are not believable will not be used. Decisionmakers are not all the same: some prefer in-depth, holistic descriptions; some value a few telling statistics; others like to see large numbers manipulated through statistical tests and use of computers. Credibility of design and methods is established by involving stakeholders at the outset in decisions about which methods to use and by discovering their perceptions of the worth of different methods. Every suggestion of new information to be collected should be scrutinized to determine how the information is directly relevant to solving the problem. When users are themselves involved in data collection, information obviously has greater credibility than when it is collected by others.

Trustworthiness

The perceived credibility of the research is intrinsically bound to the perceived trustworthiness of the researcher. If agency decisionmakers, managers or community people do not trust the researcher, they are unlikely to accept that person's judgement or technical competence. Personal rapport, flexible attitude, open mind and tolerance for high levels of frustration (it often takes longer, for instance, to reach consensus) become important characteristics for researchers who seek to be considered trustworthy. Trustworthiness is increased when participatory researchers are honest about the limitations and reliability of their findings and establish two-way relationships with managers.

Relevance

Within the time and resource constraints imposed on every study, the participatory researcher helps develop a consensus about the relevance of different types of information. The more users are involved in the research process, the more they understand it; as a result, they become interested, believe in their findings, and are likely to utilize them.

Feasibility

People have no problem in coming up with long lists of desirable and "interesting" information. The decision on what information should be gathered should be guided by relevance and priority of information to the problem to be solved. Feasibility about the process should be viewed from multiple perspectives: social, cultural and technical feasibility given time and resource constraints.

Participatory Research Emphasizes:

- Problem solving
- Expertise of the nonexpert
- Short cut methods
- Validity confirmed by consensus
- Reliability achieved by multiple methods
- Collaboration and shared decisionmaking
- Utilization of results
- Capacity building

Advantages

Participatory research has a number of important advantages over conventional social science research.

1. The emphasis on involving stakeholders and decisionmakers nurtures and builds commitment to the research process and its results. This increases the likelihood that the results will be utilized in a variety of ways at different levels, will have an impact and will bring about change.

2. When people are involved in decisionmaking, they become emotionally engaged. This releases much creative energy and supports the research process in being lively and even fun. This, in turn, reinforces involvement.

3. It results in capacity building, both within agencies and local communities. It enables participants to understand and utilize the research process to solve other problems in other contexts. Managers who are better able to understand and interpret research results are better users of research in other programs. At the community level, people who have the tools to generate knowledge in one context can gain the power to organize and initiate change in another context.

4. Participatory research results in learning, new knowledge and changes in people's "cognitive maps".

5. It gains strength from the pooling together of different perspectives, insights, knowledge and expertise to study a problem holistically.

6. It ensures that researchers stay grounded in reality and adapt the tools of their trade to the cultural, institutional and political realities of a situation.

7. It is responsive to the needs of a situation since its mandate is to solve problems by using any appropriate methodology or technique. It can borrow techniques from various disciplines, either quantitative or qualitative. When existing methods do not fit a particular situation, participatory researchers create new techniques or adapt old ones.

Disadvantages

Participatory research also has several disadvantages:

1. There can never be a "cookbook" for participatory research. The principles of participatory research can be clarified and

---

Field Insight: Ambivalence About Participatory Approaches

Many experts, technical specialists, administrators and social scientists initially may resist the use of participatory techniques at the community level. Reasons for this include:

- They, themselves, may have been trained in a highly structured information-loaded, top-down mode. Hence, they feel uncomfortable with a process that is flexible, unpredictable and involves people in the making of decisions which may be challenged by peers;

- They may feel that participatory processes, while valid, take too much time, and that better and faster results can be achieved by giving people specific instructions or messages;

- They are convinced that community-based planning and research is beyond the capacity of villagers, many of whom have little or no schooling. To devolve decisionmaking and control to the community level is therefore considered unrealistic and a waste of time;

- They feel participatory approaches confuse communities, who would prefer to be told what to do by those in authority and those who know, rather than spend time problem solving.

These and similar fears are likely to arise when people are not exposed to an experiential learning mode or training. Involvement in participatory workshops, experience with participatory tools and overnight stays in communities with skilled facilitators all help in creating openings for participatory approaches.

Source: *condensed from Lyra Srinivasan. 1992. Options for Educators. A monograph for decisionmakers on alternative participatory strategies. PACT/CDS.*

guidelines developed, but the success of the research depends on individual researchers and their ability to choose and adapt strategies and develop situation-specific methodologies.

2. Participatory research is initially more time-consuming when roles and expectations of different players are still being explored and defined.

3. It can be emotionally draining. Participatory researchers have a facilitating role — to help participatory processes that release creativity and emotional energy and feed them into productive channels. This facilitative role is not only difficult, but can be extremely tiring.

4. Conflicting data interests, value systems and data needs may be difficult to resolve to the satisfaction of all.

5. Participatory research places many demands on the researcher. Besides being a good technician, the researcher must have strong interpersonal communication skills and be adept in facilitation and negotiation.

6. Participatory research methods may not be understood by those not involved in the process and hence risk lacking credibility in their eyes.

Roles and characteristics of a participatory researcher

There are many important differences between the role of a conventional social science researcher and a participatory researcher.

Compared to engineers and technology specialists, social scientists have made a late entry into the field of water supply and sanitation. Only recently have they become involved in the front end of programs and projects on technical teams concerned with design, implementation, monitoring and evaluation of water and sanitation programs.

At a basic level, social scientists (anthropologists, sociologists, psychologists, community organizers, public health specialists) involved in project research are themselves primarily technicians. They formulate and complete the research and make their findings available to those interested; they ensure that the methodology is technically sound and rigorous, data reliable and valid, statistical analysis robust and content analysis sound.

In contrast, participatory researchers must augment their technical skills with additional abilities that allow them to play roles essential to the participatory process. Among these roles are: creative technician, human research instrument, facilitator, trainer/enabler, communicator and advocate or activist.

Field Insight: Establishing Trust Between Villagers and Officials, Rajasthan, India

The Government of India and World Bank-financed watershed development projects in Rajasthan use participatory methods in the planning of micro-watershed development. Dramatic increases in the productivity of land have been measured within two years of the program's initiation.

At the outset of the program, villagers refused to meet with officials because of a history of mistrust. Officials then decided to hold meetings in the evenings and show films. The first movies shown were, incredibly, on wetland farming in Fiji — despite the fact that Rajasthan is an arid area. The program now has its own films and a staff of 2,000. Clear principles and phases of the participatory process have evolved from a trial and error process, although success in application is still mixed. Once some community openness and trust is established, government field maps are shown to villages, problems and the potential of land discussed. To verify, elaborate and analyze the information, local residents walk with government officials through the watershed, verify boundaries, mark common land and eroded areas, and add additional features to the map. The information jointly collected is marked on the map becomes the focal point to reach consensus on an action strategy.

Source: *Anirudh Krishna, personal communication, 1994.*

Creative technician

A participatory researcher searches for creative synergistic solutions to problems. To do so requires inventing, adapting, borrowing, and changing a wide variety of methodologies: surveys, observations, cultural immersion, role plays, games, simulations, projective and semi-projective techniques and discussions. This ability is critical because the success of the participatory researcher depends on quickly processing and presenting the information that has been gathered. This must be done in ways that are credible and accepted by decisionmakers, people in the community and other stakeholders.

Human research instrument

Participatory researchers not only encourage the participation of others in the research decisionmaking process but also become participants themselves. For project staff, this may involve participating in and monitoring staff management meetings, or participating in a day of activities with an extension worker, a driller or a village woman. For village people, this may mean self-monitoring of their daily activities or those of their neighbors. In each of these situations, the researcher is the instrument for conducting observations, and for recording and interpreting data. The credibility of the results thus becomes intertwined with the credibility and trustworthiness of the researcher.

Without a high degree of self-awareness, including awareness of motives and values, participatory researchers can misinterpret situations and quickly devolve into the role of manipulators rather than enablers.

Facilitator

Participatory researchers have the responsibility to facilitate the entire research process. Experience indicates that user involvement in decisionmaking often takes place through the forum of small group meetings and discussions. Participatory researchers must be skilled, therefore, in facilitating such group discussions, creating a congenial, supportive atmosphere, drawing out opinions and containing or redirecting conflict.

Participation not only releases creativity and learning, but also emotion, both positive and negative. This is especially likely in the early stages, as well as in the later stages of data interpretation and action. Negative emotions are more likely to surface if the group involves those who see their interests being threatened by the research process questions, methods of inquiry, analysis or interpretation. This is more likely to happen in large programs where change can affect the importance, power and status of certain people.

Trainer/enabler

The role of the participatory researcher as trainer or enabler is crucial for capacity building. By presenting research as a series of problems to be solved, starting from focusing the research questions to alternative ways of disseminating information, the participatory researcher fulfills his/her role as trainer. In addition, researchers may train in specific technical aspects of research as desired by users or participants.

Communicator

The success of a participatory researcher sometimes depends more on the ability to communicate the results of a study than on a sophistication in conventional research methodology. In fact, a review of social program evaluations over a twenty-year period revealed that the more methodologically rigorous and sophisticated the research, the less likely it was to be used. For the participatory researcher, therefore, communication skills that lead to increased probability of use are essential.

Advocate or activist

Depending on the context, participatory researchers play important roles in ensuring the utilization of results and increasing the probability of follow-up action. This may be in the form of distilling results for diverse audiences, such as village people, extension workers, policy makers; holding discussion groups to review recommendations and implement change; incorporating results into training materials for project staff; or more visible forms of political activism.

Each participatory researcher has to discover his or her comfort level for follow-up responsibility, and encourage the utilization of findings by dissemination in various forms through appropriate media.

The Making of a Successful Participatory Researcher

Participatory researchers must have a high level of tolerance for ambiguity and lack of structure. Authoritarian personalities do not make the best candidate for the participatory research approach. Since no two situations are the same, participatory researchers must be willing to take risks and venture into the unknown.

Researchers who can embrace and put into practice the following *assumptions* are more likely to succeed as participatory researchers than those who cannot.

1. Research is a learning process which must be approached with flexibility and inventiveness.

2. Researchers do not know all, nor do they necessarily know best.

3. Local people and non-researchers are intelligent, creative and have important problem-solving abilities.

4. People working within agencies and in communities function within a cultural context and not a cultural vacuum.

5. People are not empty vessels, waiting to be filled by the superior wisdom of experts.

6. Indigenous knowledge systems are critical in designing and managing successful projects.

Conventional Versus Participatory Research

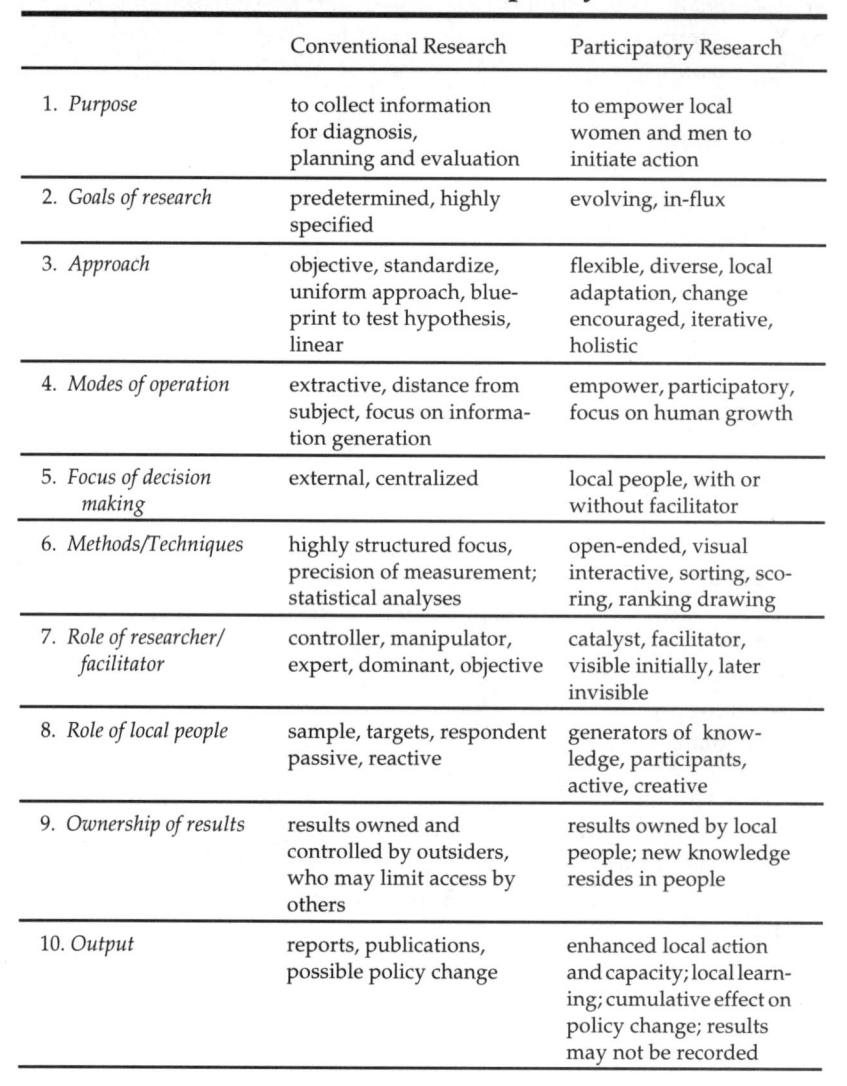

		Conventional Research	Participatory Research
1.	*Purpose*	to collect information for diagnosis, planning and evaluation	to empower local women and men to initiate action
2.	*Goals of research*	predetermined, highly specified	evolving, in-flux
3.	*Approach*	objective, standardize, uniform approach, blueprint to test hypothesis, linear	flexible, diverse, local adaptation, change encouraged, iterative, holistic
4.	*Modes of operation*	extractive, distance from subject, focus on information generation	empower, participatory, focus on human growth
5.	*Focus of decision making*	external, centralized	local people, with or without facilitator
6.	*Methods/Techniques*	highly structured focus, precision of measurement; statistical analyses	open-ended, visual interactive, sorting, scoring, ranking drawing
7.	*Role of researcher/ facilitator*	controller, manipulator, expert, dominant, objective	catalyst, facilitator, visible initially, later invisible
8.	*Role of local people*	sample, targets, respondent passive, reactive	generators of knowledge, participants, active, creative
9.	*Ownership of results*	results owned and controlled by outsiders, who may limit access by others	results owned by local people; new knowledge resides in people
10.	*Output*	reports, publications, possible policy change	enhanced local action and capacity; local learning; cumulative effect on policy change; results may not be recorded

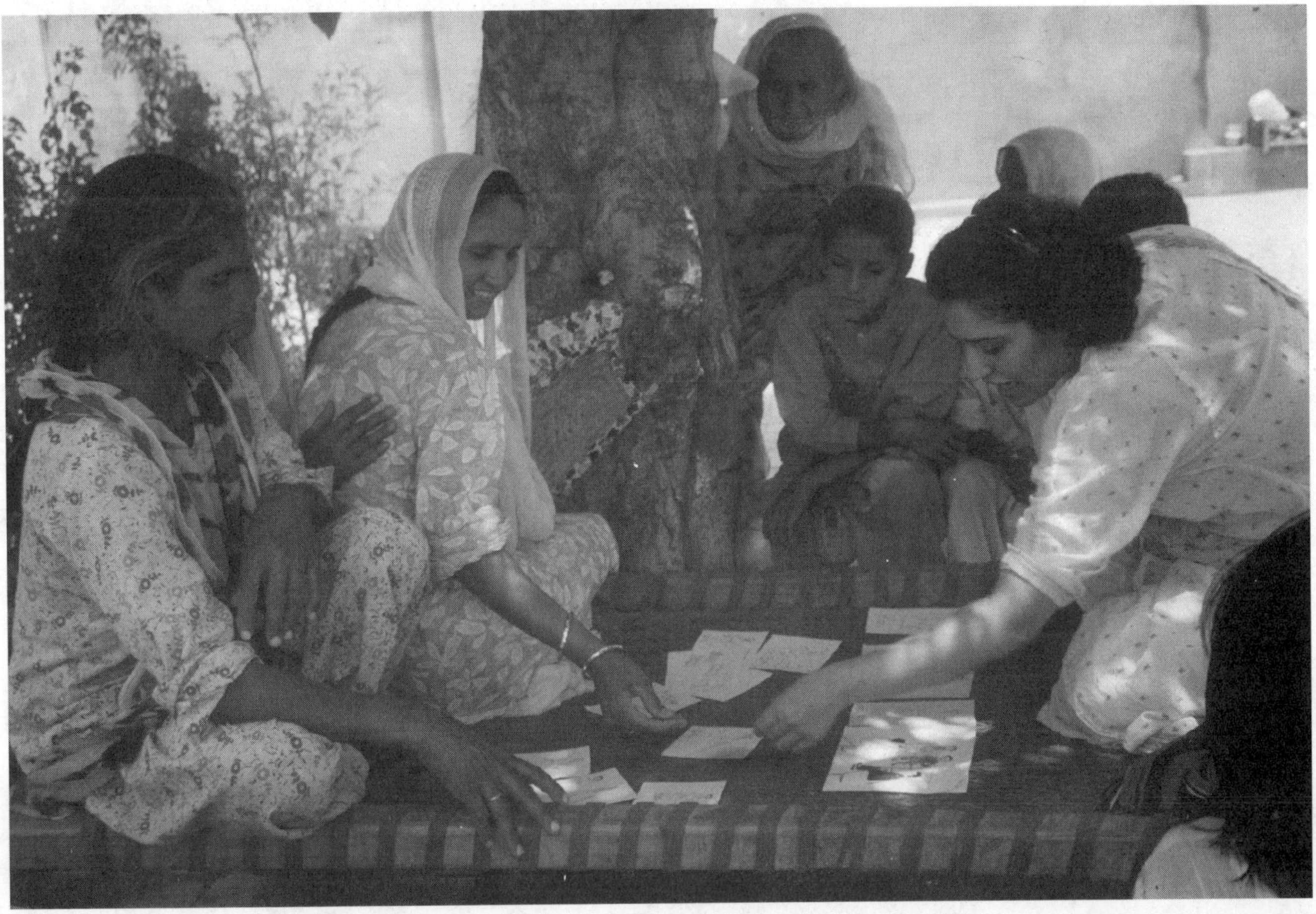

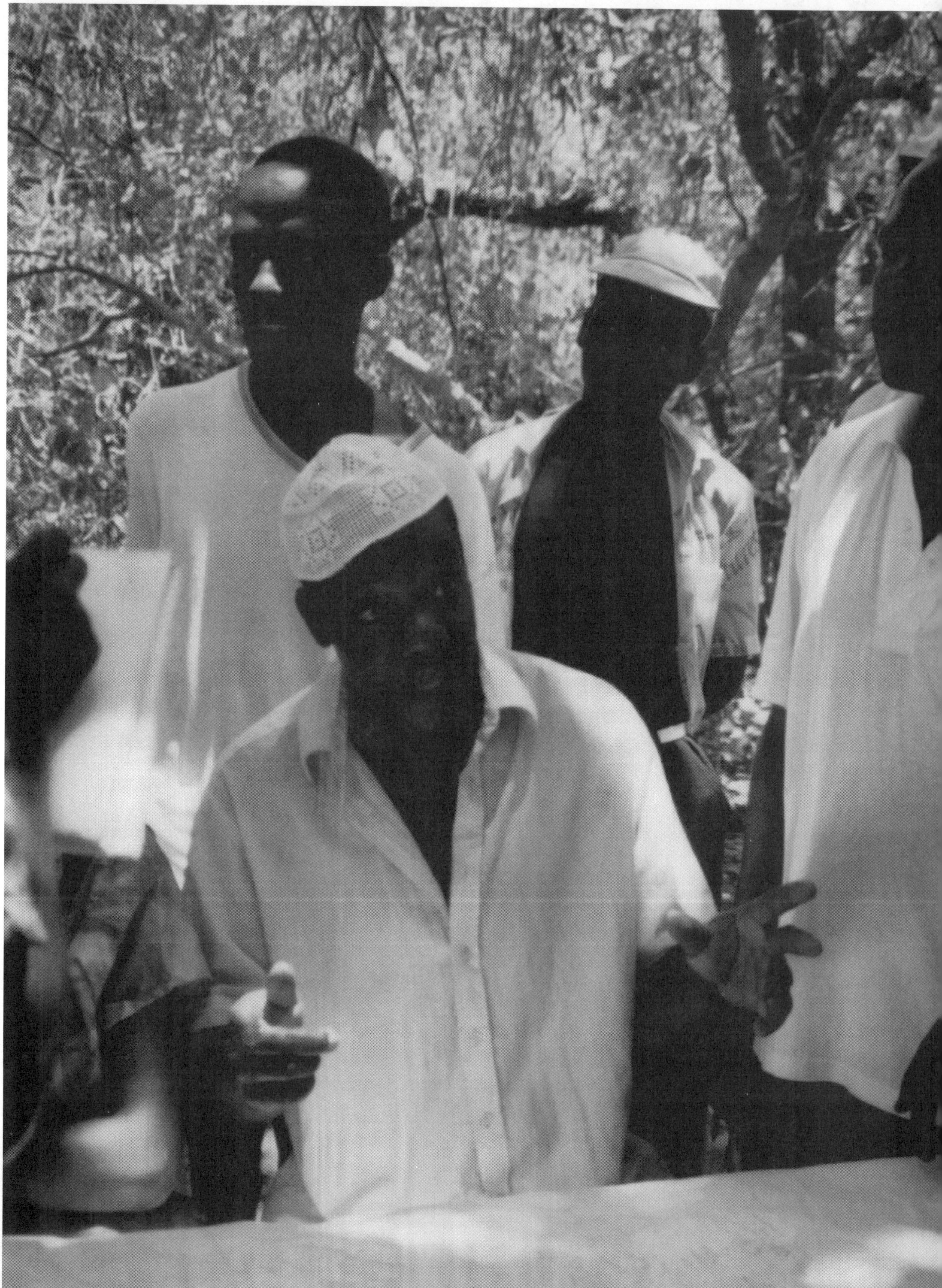

Chapter 3

Defining the Purpose of the Study

Before initiating any study, it is essential to define the goal and objectives of the entire process. If a key goal is community or staff empowerment and action, then the study should be conducted in participatory ways. The purpose of a study can be narrowly defined for a single purpose or it can be broadly defined to serve multiple objectives.

Large studies designed to gather in-depth information on a range of topics for multiple purposes (such as both planning and evaluation) can become very complex and time consuming. It is here that the constraints within which the study must be conducted and the timeframe within which results must be made available to users in appropriate forms, become crucial.

Take, for example, a study designed to collect comprehensive, in-depth information on types of water sources, their seasonal variations, their use, reliability, convenience, history and management. Information may also be needed on the presence of latrines, their use and people's felt needs for them. To make the study more manageable, if it is known that implementation of the latrine program will not begin for two years, then the research focus can be placed on studying the water supply in depth, while providing only an overview of the sanitation situation. Detailed and more timely information on latrines could be gathered before the latrine components of the program are implemented two years later. In general, small and timely studies are preferable to single large studies which attempt to understand every conceivable component of the water supply and sanitation environment in a community.

Purposes and objectives

Social assessment in water supply and sanitation projects can be conducted for a variety of purposes and objectives at any stage of a project cycle. Broadly, these include project identification, project and program formulation, assessing project feasibility, monitoring and evaluation, and training.

Four sociological factors are of particular importance and should be kept in mind when formulating any study. (See box on following page.)

Chapter Contents

This chapter explores the following issues:

- Purposes and objectives of study
  - Project identification
  - Project and program formulation
  - Assessing project feasibility
  - Monitoring
  - Evaluation
  - Training
- Methods for clarifying the research purpose
  - Involve the stakeholders
  - Make the connection between purpose and results
  - Turn the study "upside down"
  - Involve others
  - Ask for help
  - Focus on how results will be utilized and who will utilize them
- Setting limits and defining the scope of the study
  - Personnel limits
  - Financial limits
  - Time limits
- Narrowing the scope of the study
  - Utilize available information
  - Prioritize information needs
  - Degree of detail desired
  - Degree of accuracy needed
  - Checklists

> **Sociological Factors Influencing All Research**
>
> In formulating any study, researchers should keep in mind the following four factors:
>
> **Communities are not homogenous entities.**
>
> It should be assumed that priority felt needs vary among different groups in a community (social, ethnic, religious, political, age and gender groups) with the probability of income and gender differentials.
>
> **Opportunities and constraints: access and control over resources and benefits are gender specific.**
>
> Every society has gender-defined roles and opportunities which specify the relative power and spheres of influence of men and women. These need to be understood so that the data collection process taps the perspectives of and empowers both women and men.
>
> **Every society has indigenous social groupings, networks and organizations which are central to the functioning of the community.**
>
> Every community consists of informal and indigenous organizational forms which have evolved over time to ensure the smooth functioning of society. These may not be visible to the outsider because they are embedded in the social and cultural traditions of groups. Local organizational forms must be understood if external agencies (government or nongovernmental) aim to stimulate collective action.
>
> **Service delivery agencies are also social and cultural organizations with their own goals, values, rules and ways of functioning.**
>
> Every agency is composed of people in a defined, usually hierarchical, relationship with one another. Every agency has its own organizational culture and rules and regulations which define what it does and how well it does it. It is important to understand not only the technical capacity of the agency but whether the agency has the capacity and motivation to adopt a demand orientation and a community empowerment approach. For example, in a demand-based approach, agencies do not assume a high felt need in communities based on objective physical criteria such as water scarcity and distance to safe and reliable water. Instead, they focus on understanding community felt needs and willingness to pay for different service levels.

Project identification

Participatory approaches are useful in project development if those involved have the power to initiate or to influence project formulation. Thus, at the project identification stage, it may be more important to include senior government decisionmakers in a participatory process. As part of social assessment, consultative processes are particularly useful in gauging demand for the project among potential clients. A study that fails to assess a community's interest in getting involved in water or sanitation improvements often leads to lack of support during implementation. One sample Terms of Reference (TOR) for beneficiary consultation during project identification is attached as Annex 1.

Social assessment can be utilized to identify priority project areas based upon the perceived needs, demand or priority problems identified by various user groups in a community. For example, in a community in which only 5 percent of households had private toilets, a needs assessment survey found that appropriate sanitation facilities were a priority for just 5 percent of the population. The vast majority of the community simply was simply not demanding more toilets. In this environment, it would not have been feasible to initiate a toilet-construction project based on demand.

Assessing project feasibility

If plans for a project already exist, a study can be used to determine whether the plans are viable, suitable, practicable and feasible.

A feasibility study is especially appropriate when:

- There has been a long gap between any two stages of a program cycle, for instance between project identification and formulation, or between project formulation and implementation;

- Plans need alteration prior to or during implementation;

- There are periods of radical change (economic, political or policy).

For example, if a project has assumed that community contributions will be 75 percent of total cost, and the method of collection required posting agency staff in the communities for extended periods of time, this is probably not feasible. A study may be undertaken to assess the feasibility of the existing plan and explore other, more viable, strategies which put the responsibility for collecting the funds on the communities themselves, rather than on agency staff.

Project and program formulation

Formulation of a project or program requires more detailed information that is typically available from studies undertaken for the

> **Field Insight: Giving Priority to Community Information Needs, Zambia**
>
> In Zambia, the staff of a Department of Water Affairs water and sanitation program wanted to discover to what extent villagers were using an improved well. Although a formal survey had indicated that the improved well was being utilized by the majority of community members, casual observation suggested that use of the near-by traditional water source was still quite common. When the idea of a study was discussed with community leaders, they identified other priority issues which needed to be resolved, such as the maintenance of the irrigation canal.
>
> After much debate, field workers decided to utilize a community mapping process to address both the community's and their own priorities. While the exercise focused mainly on the community's priorities, it also yielded important information for the researchers on the use of the improved well.
>
> During a lengthy session that extended over two visits, community members first identified people whose plots of land directly benefitted from the irrigation canal. Then, through discussion, the villagers evolved their own strategy for addressing the maintenance problem. The mapping activity not only helped to resolve the issue of canal maintenance, but also revealed the large number of people who continued to use the traditional water source and led to a candid discussion of their reasons for doing so.
>
> Source: *Ron Sawyer, 1993, internal memo, World Bank.*

purpose of project identification. Participatory research can be carried out to assist overall program development and formulation; it can also assist in designing specific components of programs such as the management structure for the implementation phase, alternative toilet designs, methods for reaching women, and development of hygiene education programs.

Stakeholder workshops are being held with relevant municipal councils in the formulation of a Bank urban water and sanitation program in Zambia. The workshops focus on reaching consensus on the role of the municipal councils and exploring a range of delivery alternatives, including mechanisms for direct disbursement of funds to communities.

Some projects now build in pilot projects as an integral part of project preparation. These efforts can be carefully evaluated, using qualitative, quantitative and participatory methods. The JAKPAS pilot project in Nepal is one such example testing new institutional models. It includes evaluation of the performance of thirty NGOs, as well as community self-evaluation. The results will feed into refining the design of the larger World Bank-financed rural water project under preparation.

Monitoring

Monitoring allows for corrective action to be taken in an ongoing program. By involving users in monitoring activities, projects benefit from a wealth of information and observation that may be unavailable to the outsider. Through the participatory approach, community members become actively involved throughout the entire monitoring process. As a result, corrective actions can often be initiated directly and promptly; the monitoring process itself contributes to the building of local capacity for decisionmaking and community-centered development.

In Sri Lanka, in a Bank-financed rural water supply project, community involvement in monitoring is changing accountability of support organizations downward to communities rather than upward in the government hierarchy.

Monitoring by program staff is also important for providing project managers with feedback. Even in monitoring visits, it is important to be clear on the purpose. This is particularly true if any form filing is required by agency personnel. The general experience is that it is easy to collect too much information that is not used.

Evaluation

Evaluation is important in assessing if stated objectives are being met, determining factors influencing attainment of these objectives, documenting "lessons learned," clarifying follow-up actions to be taken, and measuring the impact of projects. Evaluations can be done formally or informally at any point after the start of a program.

In one project, for example, the factors affecting the performance of drillers and drilling rigs was evaluated. In a sample of twenty drillers, it was found that the most important factor affecting productivity was the lack of incentives to drill efficiently. As a result of the study, carefully structured bonuses were introduced; consequently, output more than doubled within three months.

Training

Participatory research can be extremely useful in pinpointing training needs and guiding the development of educational materials in a project.

For example, in one project, a participatory workshop involving workers from the district, sub-district and village levels was held to identify training needs and to develop materials that would assist community workers to carry out their tasks. Bringing together workers from different levels was important because each level was responsible for training the one below. During the workshop, participants wrote down their own job description and their perception of the job descriptions of others. No two job descriptions were found to be the same. The resulting discussion first focused on clarifying roles until agreement was reached. Participants then evaluated their individual abilities to carry out the various tasks, their strengths and weakness and priority areas in which they needed help. By allowing the participants to clarify their own roles and assess their own weaknesses and strengths, the workshop succeeded in developing a training program that was vital to the successful implementation of the project. (For another variation of a role perception activity, see Section II.)

Methods for clarifying the research purpose

It is not unusual to approach a study with vague ideas about what should be investigated and what the research should accomplish. However, it is absolutely imperative that clarity is obtained on the purpose of the study before proceeding.

Here are some methods that have proven useful in a variety of settings:

Involve the stakeholders

Those affected by a project — the stakeholders — should be involved in identifying the issues to be studied.

Make the connection between purpose and results

There is a vital connection between the purpose of a study, the information needed and the results that will be produced. Stating the general purpose of the research via a series of answers to targeted questions may help clarify and eventually narrow the scope of work.

Field Insight: Stakeholders Identify Need and TOR for Study, Hubei, China

In coordination with a World Bank identification mission, the East Asia Regional Water and Sanitation Group of the UNDP-World Bank Water and Sanitation Program together with Hubei provincial and city authorities, facilitated a workshop focusing on information exchange and discussion of options to improve night soil management. Participants included senior officials from the Provincial Environmental Bureau, three city environmental protection, sanitation and administration bureaus and World Bank staff.

The workshop was preceded by officials visiting different cities to study a variety of sanitation options in practice. After additional presentations and slide shows on technology options during the workshop, officials evaluated the options based on technical, financial, health and institutional requirements criteria. This very participatory process led the officials to conclude that they had insufficient information on several aspects of the different technological options.

To enable them to make informed decisions, the officials agreed that further research was needed on specific topics. Accordingly, the participants designed a research study, to answer the questions of interest to them. The results of the study will also feed into the design of a World Bank financed loan.

Source: *Workshop Proceedings, Wuhan City, 1993.*

For example, if the objective of a study is to obtain an overview of village communities and their existing water supply situation, questions such as the following may be posed:

- What is the physical area and the population of the villages?

- How accessible are villages by roads or by boats (if located on islands) in all weather conditions throughout the year?

- What is the development level of the villages?

- What are people's means of livelihood?

- How prosperous or wealthy are households?

- What proportion of households are poor?

- What are the available health, education and other infrastructure?

- What government or private development activities are ongoing?

In relation to water, the list of questions might include:

- How many water sources are there in a village?

- What types of water sources exist?

- How are they distributed spatially?

- What are the seasonal variations in water quantity and quality?

- Are the existing sources adequate in terms of reliability, convenience and access (social and physical)?

This list of questions can be expanded to yield greater detail and specificity.

A study focusing on agencies might explore:

- What is the role of communities in existing public sector programs?

- What changes in roles are appropriate among agencies? Among communities?

- What is the best combination of agencies to support a community-based water supply?

- What are their different roles, responsibilities, and incentives to perform?

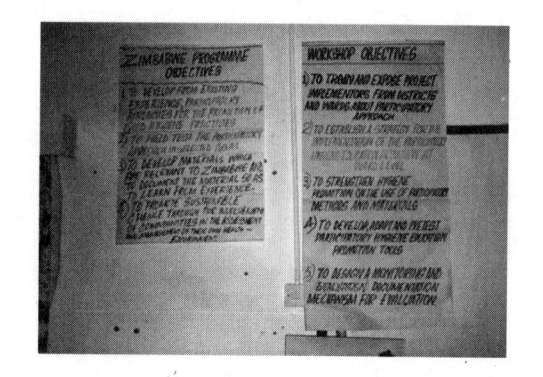

- What is the mandate and capacity of existing public sector agencies?

- What is the capacity and history of interaction between NGOs, the private sector, and government agencies?

Turn the study "upside down"

After a list of specific questions is generated, "turn the study on its head" in order to clarify its purpose. That is, instead of focusing solely on the purpose, it may prove productive to focus on the study's end product or results. By exploring the results that are to be produced, a series of questions may help clarify and make more accessible the broad, general statement of purpose. In addition, it starts the process of cutting out questions that are unnecessary to the central purpose of the study.

Helpful questions to get at the end product of the study include "What type of information is expected or desired at the end of the study?" and "How will this information be used?"

Thinking about the research end point may reveal that much socioeconomic information that could be gathered, such as household composition and income levels, is not what is really needed. What may prove necessary, instead, is employment of easy-to-rate general indicators such as accessibility of the village, already-existing infrastructure, people's experience with technology, existing water sources (their type, accessibility, and reliability). It may also be important to find out who in the community perceives water as a problem and who does not.

Through this approach, the original objective identified above — "to obtain an overview of villages and their existing water supply situation" — can be refined to read:

- To obtain an overview of the development level of a village;

- To understand the existing water supply system, especially with reference to functioning, access and use; and

- To assess whether water is a perceived priority problem and whether there is a demand for improved supply.

These refinements allow for a much more focused, and ultimately useful, data collection process.

Involve others

No successful water supply or sanitation program at any level can be a one person operation. Project colleagues, both those who are more senior and those who are more junior or field-based, are repositories of valuable information that has been developed through years of hard-won experience. Because technical people are generally not prolific report writers, and community workers never seem to have sufficient time to record their observations, their insights must be drawn from them, using basic techniques such as informal meetings, brief workshops or through a simple questionnaire. Gaining access to their knowledge will enhance understanding of issues that should be addressed in a study, and will further clarify research objectives.

Ask for help - brainstorm

People are sometimes reluctant to ask others for their ideas and experiences because they fear that soliciting help to define a problem reflects poorly on their competence. On the contrary, people are usually pleased to be asked for their assistance. Doing so not only makes a study more useful, it also generates support, interest and the beginning of a commitment to the research findings and the undertaking itself.

In a small island nation, for instance, the head of a water supply and sanitation section wanted to undertake a study to find out about the use and status of latrines and sanitation in remote islands. Interministerial cooperation was essential but difficult to elicit. Discussing the research and seeking ideas from senior people in relevant ministries not only assisted in clarifying the issues to be addressed, but provided valuable assistance later in implementing the study and ensuring that the results were used in formulating a national program.

In asking for help, one caveat should be noted, however. A brainstorming session involving senior and junior staff may not be successful if it takes place in a formal and hierarchical setting. In fact it may be counter-productive if the session becomes one of pronouncements from senior people rather than one of dialogue. This may create a situation in which the person in charge of the study feels compelled to do as senior officials have publicly advocated for reasons of deference. In this type of situation, one-to-one informal meetings may be more appropriate.

Focus on how results will be utilized and who will utilize them

Focusing on how results will be utilized, who will utilize them, and at what levels they will be utilized can help to clarify what questions should be asked and in what detail. Will the research be used for planning, feasibility, assessment or training? Knowing this will help in defining specific purposes.

A study that has been designed to gain information for project identification will have a different level of precision, degree of detail and type of information needed than will research conducted for evaluative purposes.

The degree of specificity of information will also be determined by the level at which planning is carried out. The closer the planners are to the level of implementation, the more the degree of detail of information increases. Thus, planners at the sub-district level will need a more detailed overview of specific villages than those at district, provincial or state levels.

Setting limits and defining the scope of the study

Once the purpose, anticipated results, and intended use of results are clarified, the limits and scope of the study should be set. The tendency in most studies is to become too ambitious and collect all kinds of information, some of which is important, some less so and some merely interesting but of no relevance to the project at hand.

To avoid spending time and resources on developing unnecessary information, it is important to consider the purpose and objectives with reference to *constraints* within which the research must be conducted.

Constraints fall into three categories: *personnel limits, financial limits and time limits*. Once again, the simplest way to understand the implications of various constraints for the scope of the study is to pose a series of questions.

Personnel limits

- Who will conduct the study?
- If it is the project planner, will he/she concentrate primarily on the study for a defined period, or will it be conducted in addition to his/her other duties?
- Will an outsider be commissioned to do the study? What will his/her skills be?
- What resource people will be available to assist with the planning of the study?
- How much time can they be expected to devote to the study?
- What personnel are available to collect data; to do the field work in data collection?
- What personnel are available to do data analysis?
- Will computer analysis be necessary?
- Are people available to analyze data utilizing the computer?
- What media will be used to disseminate the findings?
- Who will write the report or make findings available to different categories of users?
- Who will need to make subsequent project decisions?
- Will they be involved in design, implementation, discussion of results and specific planning of steps for follow-up?

Financial limits

- How much money is available for different phases of the study?
- What facilities and materials are available for the different phases of the study? These phases include:

Field Insight: Broad Surveys Yield Unhelpful Results, Karnataka, India

In the pre-planning phase of the government of Karnataka water and sanitation project financed by the World Bank, a local consulting firm was hired to implement a series of social surveys on broadly defined issues such as health, sanitation, and water. The researchers used a variety of techniques, both participatory and traditional, to complete the study in about thirty villages. They conducted surveys, focus groups, and interviews, and did participant observation, staying in the villages overnight and eating with members of the community.

The detailed results of the study based on these social surveys did not prove particularly useful because the surveys tried to cover too many topics and were too general. However, there were a number of important side-effects from the demonstration of Participatory Rapid Assessment (PRA) methods. First, PRA helped convince the Bank staff that participatory methods were important and should be incorporated in the main project. Second, the relationship forged between the NGOs and community members proved invaluable in the implementation phase. Finally, because Bank and government officials were convinced of the utility of the PRA methods, it became easier to convince the initially resistant project engineers of the importance of incorporating villagers' decisions into the design.

Source: *Ellen Schaengold, personal communication, World Bank. 1994.*

- Planning
- Training of field workers
- Development of data collection techniques
- Data collection-field work
- Data processing and analysis
- Report writing and other techniques of data dissemination
- Discussion and dissemination of results to plan follow up

Time limits

- What is the time framework for the study?
- When will the study be initiated?
- How much time can be spent in the field?
- How much is needed for data analysis?
- By when should results be available and in what form to different categories of users?

Once these and similar questions have been answered, the realities of resources available and the time framework within which the study must be completed will be evident. In general, too much time is spent on data collection and too little time on analysis and marketing the results.

Narrowing the scope of the study

When the purpose of a study is clear and the resources available have been identified, it is important to narrow the scope of the study to manageable proportions. A research study is like a living organism: each part influences and is influenced by every other part. Thus, decisions on what specific types of information are needed, in what depth and to what degree of accuracy influence how the information will be collected and how it will be processed.

Although decisions about methods of data collection are not of concern at this stage, it is important to think about types of information, and the degree of detail and accuracy desired. It is therefore desirable to know where one is willing to make compromises and where one is not.

Strategies for narrowing the universe of potential information include utilizing information that is already available and prioritizing information needs, including the degree of detail and accuracy that are required.

Utilizing available information

Before initiating a study, the researcher should do a quick but wide-ranging search for already available information: old project documents, previous studies, evaluation reports and reviews. The best place to start is within the program offices.

It is also important to look beyond the program premises for information. It is better to discover at the outset of a study that a local NGO has been working in a similar region and has collected much information over the years than when it is time to circulate study results. Government departments likely to have information include census bureaus and departments of statistics, ministries of agriculture, health, education and the environment, land boards and water boards. In addition, NGOs, local universities, consulting firms and international organizations all may have information relevant to the project.

Once the search is completed, the existing information should be compared to the information needed. Include only what is relevant for your purpose.

Prioritizing information needs

As a decisionmaking tool, it is also helpful to prioritize in some detail the categories of information needed. Again a common sense approach works best. Simple questions may help clarify priorities: "What must I absolutely know by the end of the study?" "What can wait until another time?"

Degree of detail desired

Another important decision is the depth to which any given topic should be covered. For example, is it sufficient to know about people's primary occupation, or is information about subsidiary occupations or seasonal work needed? Similarly, if information about latrines is required, is it sufficient to know about the total number of latrines in a community, or is information needed about when they were built and how and why they were built? Or is the critical information really about ownership patterns?

Information obtained from a review of existing reports and from workshops or meetings with people may shed light on a variety of topics. This information should be examined by asking the question "Do I need additional details relevant to this topic?"

Degree of accuracy needed

A typical problem in data collection lies with the assumption that every item of information should be precise to the "nth" degree.

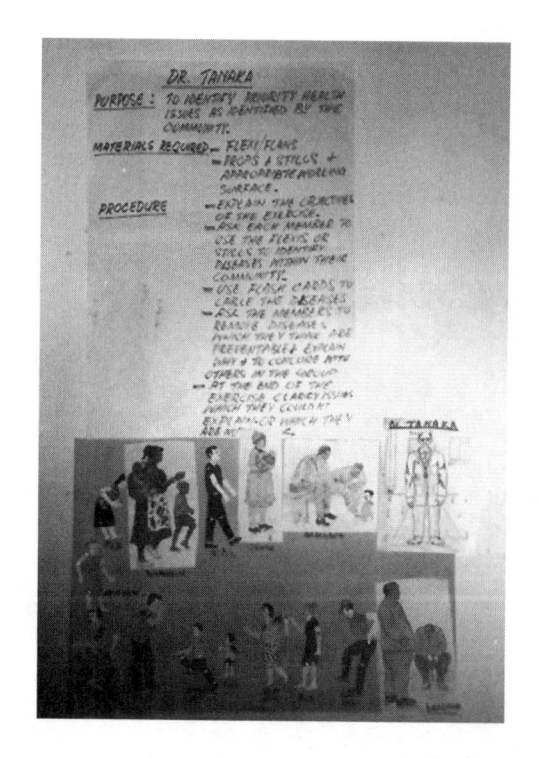

Although such precision or accuracy may be appropriate in a pure research project or academic exercise, for participatory studies, it is usually not required. An exception may be for valuative purposes in which precision may be needed to assess specific issues, such as time saved in water collection by the opening of new sources.

Time, money and headaches can be saved if the acceptable levels of accuracy for data are explored and determined prior to making a decision about methods of data collection. For example, many studies try to assess precise household income levels, even though determining income levels accurately is extremely difficult. Before embarking upon such a course, researchers should consider the relevance of such information. Could some other general indicator of wealth or availability of disposable cash be sufficient?

Studies that try to be sensitive to women's potential in water supply and sanitation often try to measure "time allocation" for various tasks in hours and minutes. Collecting this kind of data is time consuming. For project purposes, it may be enough to assess the daily and seasonal activities of the women involved.

Another example is time taken for water collection. Are broad categories of information sufficient for project purposes or is it important to know how many minutes and seconds it takes for a women to collect water?

A simple format like the one below can assist in prioritizing data in acceptable levels of detail and accuracy.

Defining the Purpose and Objectives of the Study

Purpose

- What is the purpose of the research?
- What are the primary and secondary objectives?
- Who will lead the study?
- What are the constraints—personnel, financial, time?

Results

- What types of results or information are needed?
- What are the action implications of the results?
- Who will utilize the results?
- How will results be utilized?
- When are the results needed?
- Why are the results and information needed?
- Where will the results be utilized?

Checklists

A checklist of types of information that may be needed in a water supply and sanitation program has been provided in Section III of this document.

The length of the checklist makes it obvious that a small study could not possibly answer all the items in depth within a short period of time, even within a single community. The problem becomes more complex if several communities or types of communities have to be studied or sampled at the same time. Hence, the importance of defining what information is needed, in what depth, to what degree of precision and by when.

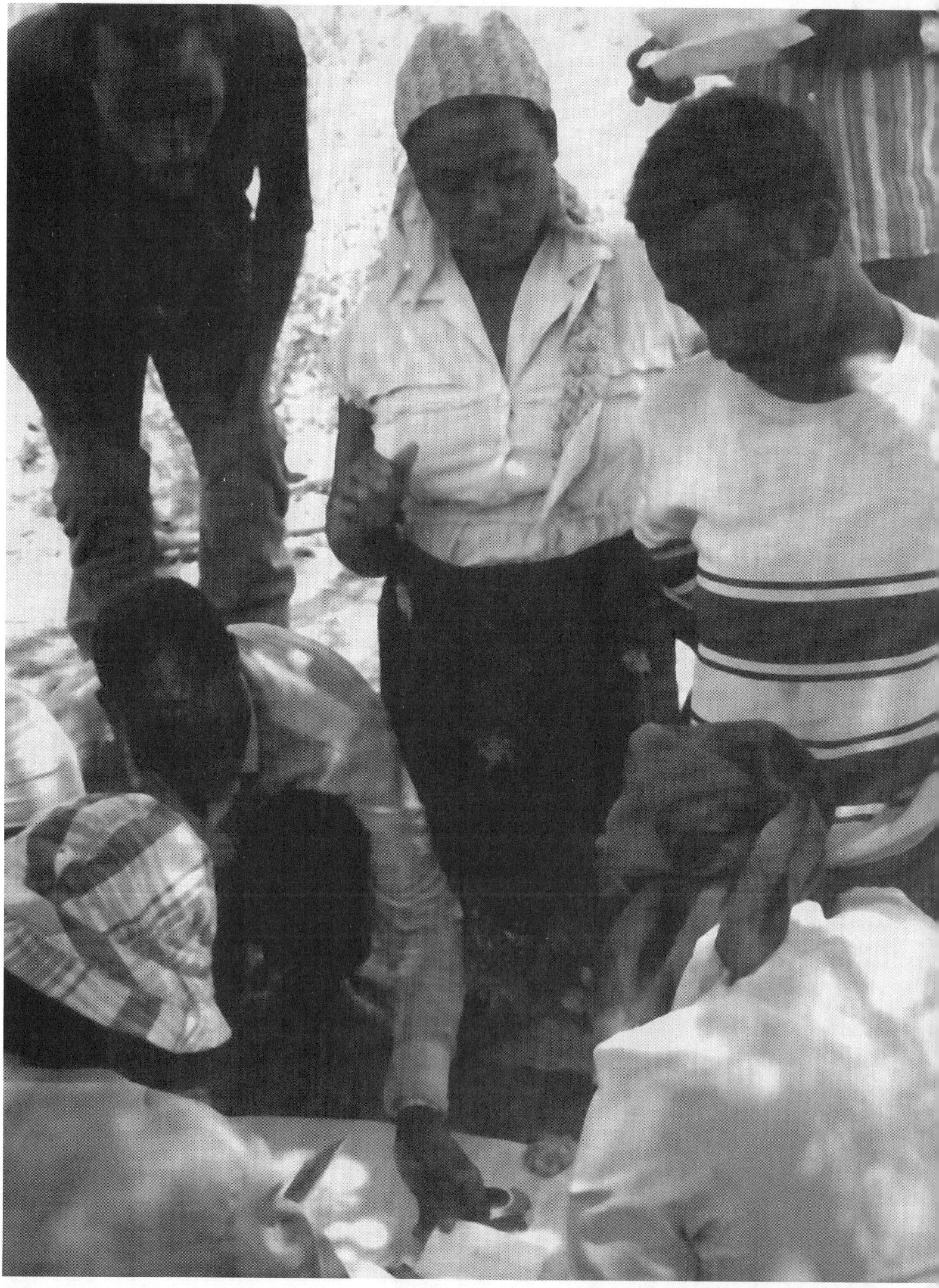

CHAPTER 4

ORGANIZING FOR THE PARTICIPATORY RESEARCH CYCLE

Stages of participatory research

In the participatory research cycle, users must be meaningfully involved at each stage, and their opinions, desires and preferences given priority at each step of the study. To achieve this, participatory data collectors place emphasis on involving users and stakeholders at both the beginning and the end of the participatory research cycle.

Involvement of users usually varies in form for agency staff and for community members. Agency staff can meet during the day, in a conference room, complete and mail back questionnaires or participate in small group discussions. Community members may participate in small group discussions at night or under a tree. Depending on the situation, local people may be more eager to participate in a survey than project staff who may view data collection as outside their job descriptions.

The ideal scenario of participatory research is presented on the following page. In reality, the process can become quite messy, the stages are less clear, and the movement of the survey tends to be cyclical rather than linear. Each stage is briefly described below.

Request for assistance or information

Participatory research usually begins with a request from the community or from an agency or project manager. Participatory research is usually facilitated by project staff who have an ongoing relationship with project management or by outsiders hired specifically to facilitate the study. As noted in Chapter 3, in responding to this request, the job of the researcher is to serve as a facilitator to assist people to understand the purpose and objectives of the study and what information is needed. As the facilitator rather than the director of the study, the researcher becomes the human instrument who supports and empowers the local community and others to design or undertake the research.

When the request for assistance comes from managers and communities involved in managing water supply and sanitation facilities, the task of doing research is simplified. But even then, the process can be a challenging one. The manager of one project, for example, wanted help to design an economic impact study of the relationship of women and water. Detailed terms of reference were

---

Chapter Contents

This chapter explores the following issues:

- Stages of participatory research
  - Request for assistance or information
  - Shared definition of research questions
  - User elaboration and prioritization of research questions
  - User definition of kinds of answers desired
  - Shared discussion and selection of methods
  - Collaboration for data collection
  - Joint data analysis
  - Joint interpretation of findings
  - Multimodal dissemination of findings
  - Action
- Institutional arrangements to support data collection
  - Private research institutes and consulting firms
  - NGOs
  - Universities
  - Individuals
  - Contractual arrangements

worked out but were not approved. A year later, when the study had still not been initiated, intensive frank discussions with the manager established that he wasn't really interested in the proposed study. His two priority needs were to ensure the project staff were competent to undertake research and evaluation activities and to develop a systematic delineation of the role of women beyond just measuring the economic impact of this activity. Once this was clear, the study was redesigned to address his concerns.

At the community level, the need for a study is often identified as a problem that needs to be resolved. For example, in community meetings in Kenya, local leaders reported a great need for handpumps, yet they had difficulty generating community interest and contributions to install them. When the same pattern appeared in several communities, project staff designed a one day workshop bringing together village chiefs to identify the key issues. The chiefs did not want to be involved in data collection because they felt local people would be intimidated by them. The actual data collection was undertaken by project staff and findings reported to the village chiefs.

If no request is forthcoming, the researcher can help set the stage for the study by using a variety of techniques such as those found in Chapter 5 to help stimulate thinking about the community or agency problems and how to address them.

Shared definition of research questions

In developing a shared definition of research questions, nothing should be taken for granted. Even though goals and objectives may be spelled out in existing project documents and written terms of reference may be available, these should not be accepted as "gospel". Experience indicates that managers often have difficulty drafting clear terms of reference.

Since participatory water and sanitation projects are dynamic and interactive, the needs of project managers or community people may be different from those developed at the time of project formulation. It may also be the case that the original project goals and objectives were not developed with full community involvement and therefore are not representative of the desires and needs of local people. In addition, the goals and objectives within a project document may be so broad that their interpretation is a matter of personal preference. Within the water and sanitation sector, an example of a common goal that is open to broad interpretation is "improved health and quality of life."

Each project or program has multiple goals, some stated, some unstated; some are clear from the outset while other emerge or crystallize with experience. Goals and objectives are sharpened by centering discussion on what personnel would like to achieve, what changes they would like to see, what their work involves and difficulties they experience. This phase is characterized by brainstorming

Stages of Participatory Research

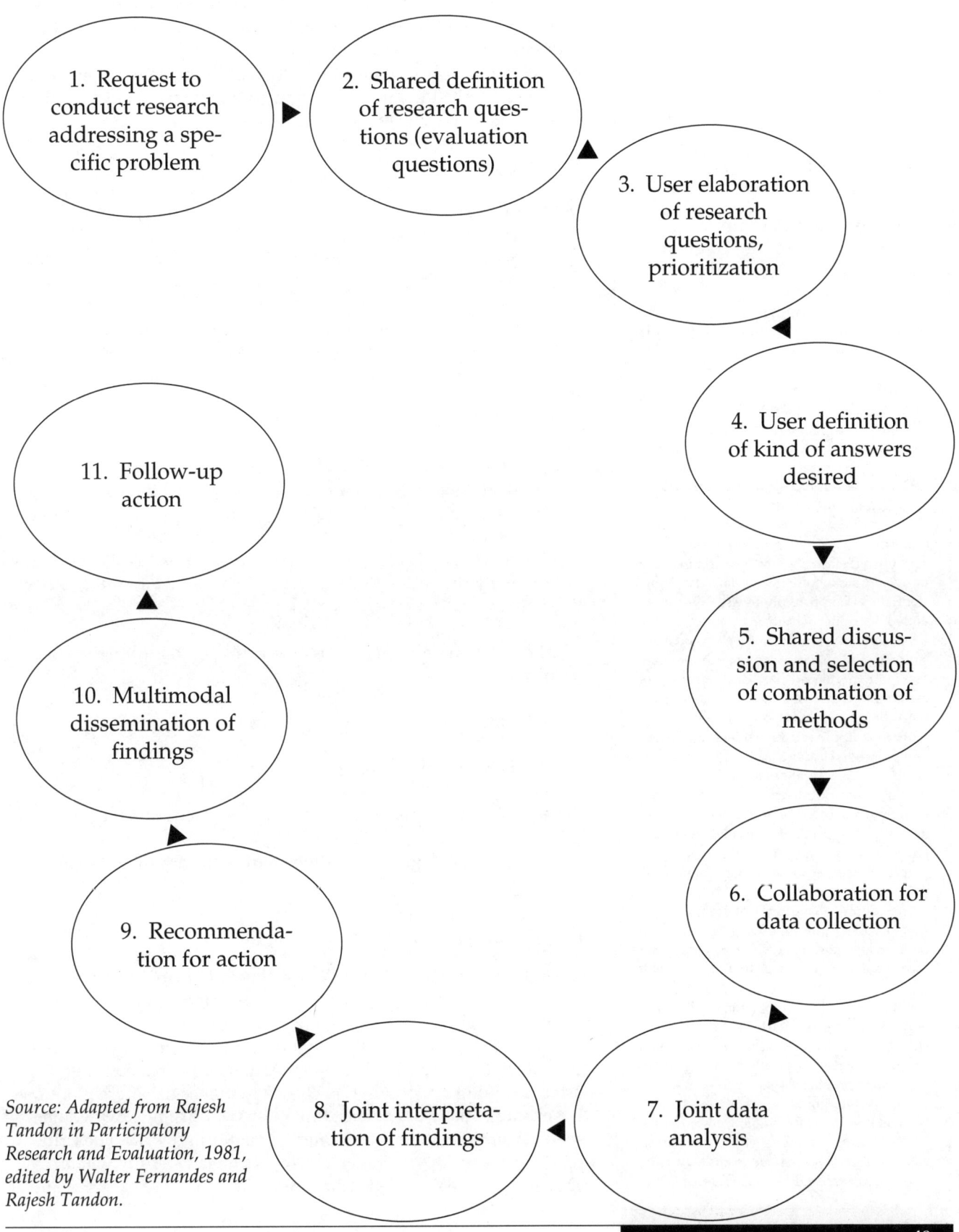

1. Request to conduct research addressing a specific problem
2. Shared definition of research questions (evaluation questions)
3. User elaboration of research questions, prioritization
4. User definition of kind of answers desired
5. Shared discussion and selection of combination of methods
6. Collaboration for data collection
7. Joint data analysis
8. Joint interpretation of findings
9. Recommendation for action
10. Multimodal dissemination of findings
11. Follow-up action

Source: Adapted from Rajesh Tandon in Participatory Research and Evaluation, 1981, edited by Walter Fernandes and Rajesh Tandon.

with users at different levels. Research questions may either be formative (helping design a component or project) or summative (assessing achievement and making recommendations on whether a project or program should continue). Applying goal-oriented project planning techniques — such as ZOPP and logframe — to the participatory research study may be helpful at this stage, in creating both greater understnding and commitment.

User elaboration of research questions, prioritization

It is better to start a study with too many questions without answers than to end up with too many answers without questions. The skilled researcher acts as a facilitator in assisting people to elaborate upon the questions, group them and establish priorities.

Research priorities should be established through discussion. This can be done in a variety of ways. A list of issues, for example, can be read and people can prioritize them by voting, or, small group discussions can be held in which people draw pictorial representations of what they know and what they need to find out.

It is rare that everyone's questions can be answered in the desired depth given the limited resources and time available. As a decisionmaking tool, it is useful to prioritize categories of research issues in some detail. Again, a common sense process works best. Simple questions may help clarify priorities: "What must absolutely be known by the end of this study?" "What can wait until another time?"

The following four questions are useful in establishing research priorities:

- How will the findings be utilized?
- How important are the findings?
- What is the relevance of findings to decisionmaking?
- To what degree do findings reduce uncertainty or increase understanding about a situation?

User definition of kinds of answers desired

A single research question can be answered in many ways: in concrete or abstract terms; in specific or general terms; with precision or approximation; in sociological, poetic, philosophical or statistical terms. Rather than second-guess which types of answers are desired and which have credibility, the researcher should spend time with users exploring options and levels of acceptance of different options. For example, there are at least three ways of learning how women spend their time: follow six women for a day and write a descriptive report; ask a group of women to use pictures to identify which activities are the most time consuming; or conduct spot observations

---

Field Insight: A Participatory Process in Developing Hygiene Education Programs

Participatory planning and stakeholder workshops have been the key elements in the successful development and implementation of a regional participatory hygiene education program supported by the RWSG in East Africa and WHO. Within two years in-country programs have been initiated in Kenya, Zimbabwe, Botswana and Uganda with support from national governments, UNICEF, FINNIDA, SIDA, DANIDA, AMREF and universities.

Since the idea of participatory hygiene education was relatively new, the first meeting of the key stakeholders was a ten day experiential workshop involving a series of open-ended learning activities to focus on the behavioral, cultural, technological and institutional correlates of hygiene education. Having experienced the participatory process rather than being told about it, participants became convinced of the value of participatory approach to hygiene education and proceeded to develop programs for their own countries. This was followed up by the participants mobilizing resources within country to put their vision into action. RWSG and WHO specialists provided technical support when requested and brought participants together to exchange experiences.

All the countries have conducted multiple program level workshops, trained local artists to develop hygiene related visual materials, and incorporated materials in training of health and community development staff. With full country ownership, every program is different and evolving through learning by doing. A strong regional network has emerged with countries increasingly calling on each other rather than on external expertise.

Source: *Condensed from Ron Sawyer, Participatory Hygiene and Sanitation Transformation: A participatory process for institutionalizing change, draft report, RWSG East Africa, 1994.*

on women's daily activities on a random sample of 100 women over a year and report the findings in tables and statistical analysis.

Shared discussion and selection of methods

The participatory researcher works with users and stakeholders to find creative and innovative solutions to problems. In determining the data collection methods to be used, an assessment of strengths and weaknesses, both technical and pragmatic, should be considered. This also includes discussion of who would collect data or train for data collection. More information on the selection and training of field workers is available in Chapter 8.

Collaboration for data collection

Because most data collection involves more than one person, it inherently requires collaboration. If data are to be collected over a period of time, then considerable organization is needed to undertake data collection successfully.

The process of organizing for collaborative data collection builds support among users and eliminates many problems commonly faced by researchers. For example, agency managers become more sympathetic to transport needs; community members involved in discussion of where the facilitator should stay are less likely to be offended by selection of particular family. When people are involved in the decision of what to observe or whom to interview, they are less likely to be offended by choices made and are more likely to ensure that short cuts in sampling do not result in a skewed sample.

Joint data analysis

For a full treatment of this stage of the research process, see Chapter 9.

Joint interpretation of findings

"Ultimately a good explanation is one which makes sense of the behavior, but then to appreciate a good explanation, one has to agree on what makes good sense." (J.C. Taylor, 1971).

Data do not mean anything until we attach meaning to them. Participatory researchers do not "search for the truth;" they know there are many truths. What a behavior or action means depends on its context and varies with the perspective of the viewer. Once again, the role of the researcher is not to be the sole proprietor for data interpretation but to work together with others to identify alternative interpretations. This process not only decreases the probability of misinterpretation, but also establishes authenticity and credibility of findings.

Multimodal dissemination of findings

Researchers have to be inventive in disseminating their findings in order to catch the attention of decisionmakers. For their part,

Field Insight: Joint Data Collection and Analysis Between Community and Engineers, India

Action Aid, an NGO involved in implementation of the government of Karnataka and the World Bank rural water and sanitation program, provides a vivid example of a participatory process which integrates the knowledge of local communities with the technical expertise of engineers.

In each village, Action Aid staff ask community members how they want the outside teams to work with them. After setting up work schedules, the community divides into smaller groups to draw detailed maps of their neighborhoods; each house is represented by a square. These maps include drawings of the drainage, water and sanitation systems and public facilities in the area. The small neighborhood maps are then transferred onto a large sheet of paper, where they are joined together to represent the entire village. Any missing details are added by community members.

The engineer's design translucent map depicting the pipe water system and drainage is then overlaid onto the community map. The engineers and community members discuss possible changes and negotiate the final version of the design. The community drawn plan is presented by the engineer to the village leaders and neighborhood representatives. Finally, this design is painted on a wall in a public place in the village. This public map serves as a monitoring tool and ensures public knowledge and accountability.

Source: *Sam Joseph and Bhakthar Solomon, personal communication, Action Aid, Karnataka, India.*

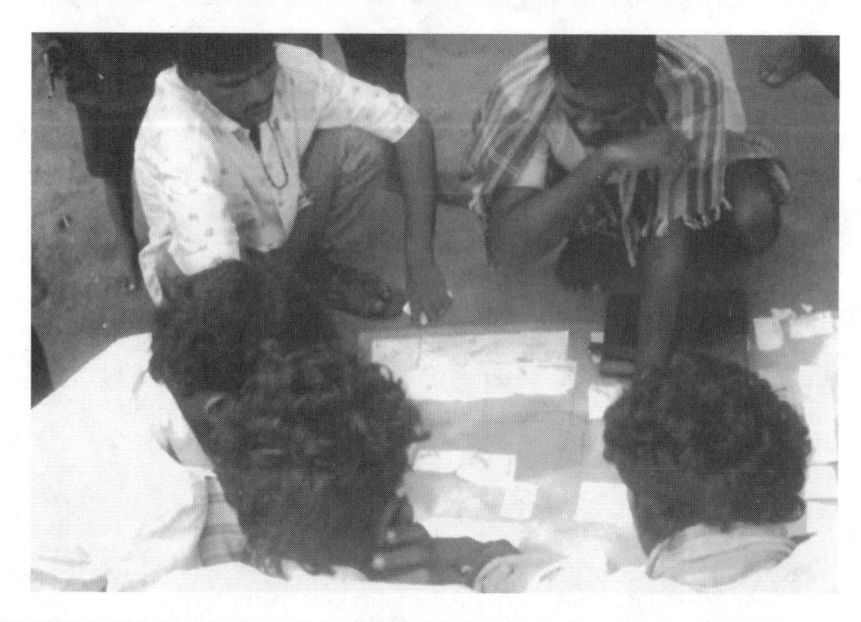

53

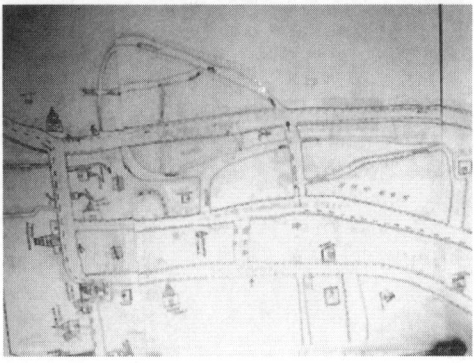

decisionmakers often need the leverage provided by research findings for justifying their actions or lending credibility to their arguments and requests. Finding the right vehicle to disseminate findings can make the difference in whether or how those findings are utilized.

If a written report is chosen as the means of dissemination, it is important to determine who will write the report, who would like to contribute, and the preferences of decisionmakers as to language, style, length, technical details, tables, maps, graphs, photographs, and descriptions. Often the most influential people do not have the time or ability to read long reports. This includes managers, policymakers, project staff and community members.

There are many ways of disseminating findings in addition to written reports; indeed, many of these alternatives can be even more effective since cognitive styles vary and people learn in different ways.

Findings can be very effectively disseminated through meetings, workshops, conferences, seminars, flyers, photo displays, slide shows, model displays, displays of data collection instruments, role playing, songs, and skits. At the community level, street theater, village meetings and mapping have all been effectively used to feed back research findings and discuss follow-up action.

Action

Participatory researchers should involve themselves in follow-up action as needed. At the very least, they have the responsibility to provide support that will help lead to action. This may take one of several forms: reorienting the training of health workers; developing training materials for sanitarians, community workers or researchers; developing ideas for scripts for slide shows; setting up a monitoring or evaluation system; training project staff in conducting participatory meetings; even assisting in the redesign of entire programs.

In the action stage, the role of the researcher will vary. However, researcher involvement in action needs to be conducted in ways that enable participation rather than create new dependencies.

Institutional arrangements

The institutional arrangement for research is a factor that is crucial in participatory data collection, but whose importance and overall influence is often overlooked. Institutional arrangements are particularly important when research is being conducted for agencies prior to project formulation.

Ideally, research should be conducted by program or project staff in collaboration with the community or agency staff. However, it is often the case that the staff lack the capacity, conviction or time to carry out a research study. When this is the case, special research components can be built into projects for specified purposes, such as design, monitoring or evaluation. Often the financial support for such research stems from external support agencies, national foundations or universities.

If the research is to be useful at the project or program levels, rather than simply serve the purpose of donors, then the issue of accountability and the definition of task for researchers becomes very important.

Several institutional arrangements have been used in projects, some of which have been more effective than others. Each has advantages and disadvantages.

Private research institutes and consulting firms

These can be very professional and efficient. If delays occur, however, or if extra work is needed for which there is no financial compensation, private sector groups are rarely willing to make the extra effort. Unless the leadership has a personal interest in the research, such institutes are unlikely to feel a personal stake in meeting clients' needs beyond those expressed in contractual arrangements. For participatory research in which the process itself is of the highest value, the situation is often less than ideal.

When studies are funded by external funds, the relationship between donors, national program agencies and research institutes is often found to be poor. This is aggravated if researchers have no direct contracts or accountability to the national project implementing agency, but rather feel accountable to distant donors. There have even been cases in which a consulting firm refused to make a copy of its final report available to the national implementing agency because the firm's contractual responsibility was to the donor. Experience shows that when contractual arrangements have specified payment conditional on satisfying the implementing agency managers, consulting firms have had a very different perception of their primary client.

NGOs

Many NGOs in developing countries are on the forefront of participatory research. However, they often have limited capacity to undertake commissioned research or even make their best participatory trainers available, especially on short notice. Given these constraints, some of the larger NGOs have created consultancy units within their organizations to facilitate the work of others without disrupting their own ongoing programs.

Universities

If suitably qualified individuals can be identified within a university, contractual arrangements can be very satisfactory. University-based researchers often have professional interests and satisfaction in conducting thorough research. In addition to bringing money into the university, applied research provides them with experience and field sites for training students. However, university professors also operate under time constraints and are limited by the dictates of the academic calendar, semesters, exams and vacations.

Each university also has its own internal rules which affect the availability of professionals and students for field work. Often, the same group of researchers is selected for externally-funded projects. Over-commitment results in their being the primary investigator in name only; consequently the quality of work suffers.

Individuals

If individuals are contracted directly by the project rather than by donors, they can give the project their individual attention. If chosen well, individuals can be dedicated professionals who enjoy contractual work and work hard for the project. However, their lack of an institutional base can be a disadvantage. If the person leaves the work incomplete, the implementing agency has little recourse.

Contractual arrangements

For participatory research, no matter what the institutional framework, research institutes must be held accountable to implementing agencies as well as funders. This is easier if research and implementation are conducted under the auspices of one organization.

Since the methodology of participatory research gives primacy to users, its methodology should become part of the contract. This should include manager involvement in identifying research priorities, agreement on timing, methods, nature of data and mechanisms to ascertain quick feedback to project staff. For example, maps of water sources and settlement patterns can be developed and handed over before detailed attitudinal data.

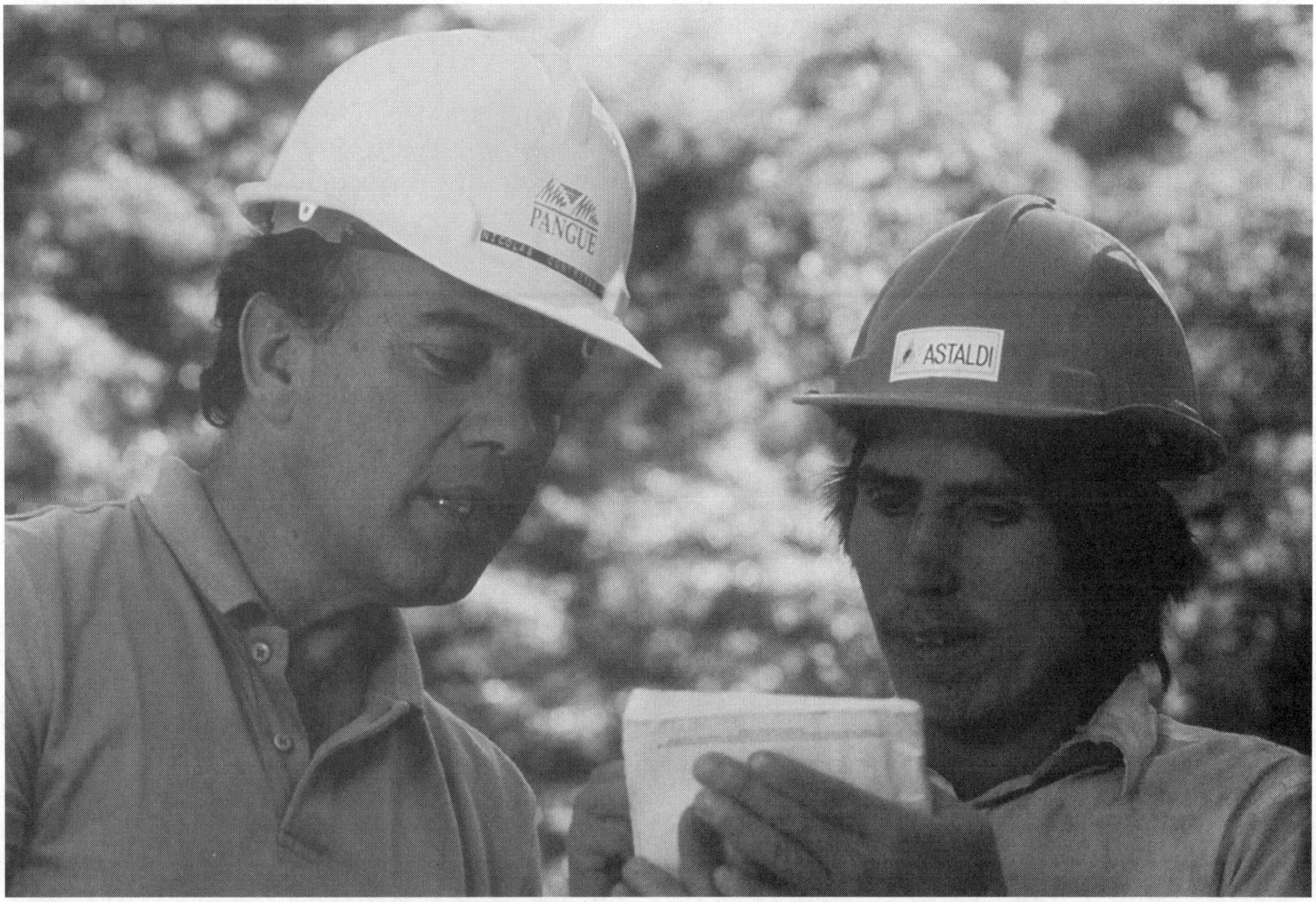
PANGUE
ASTALDI

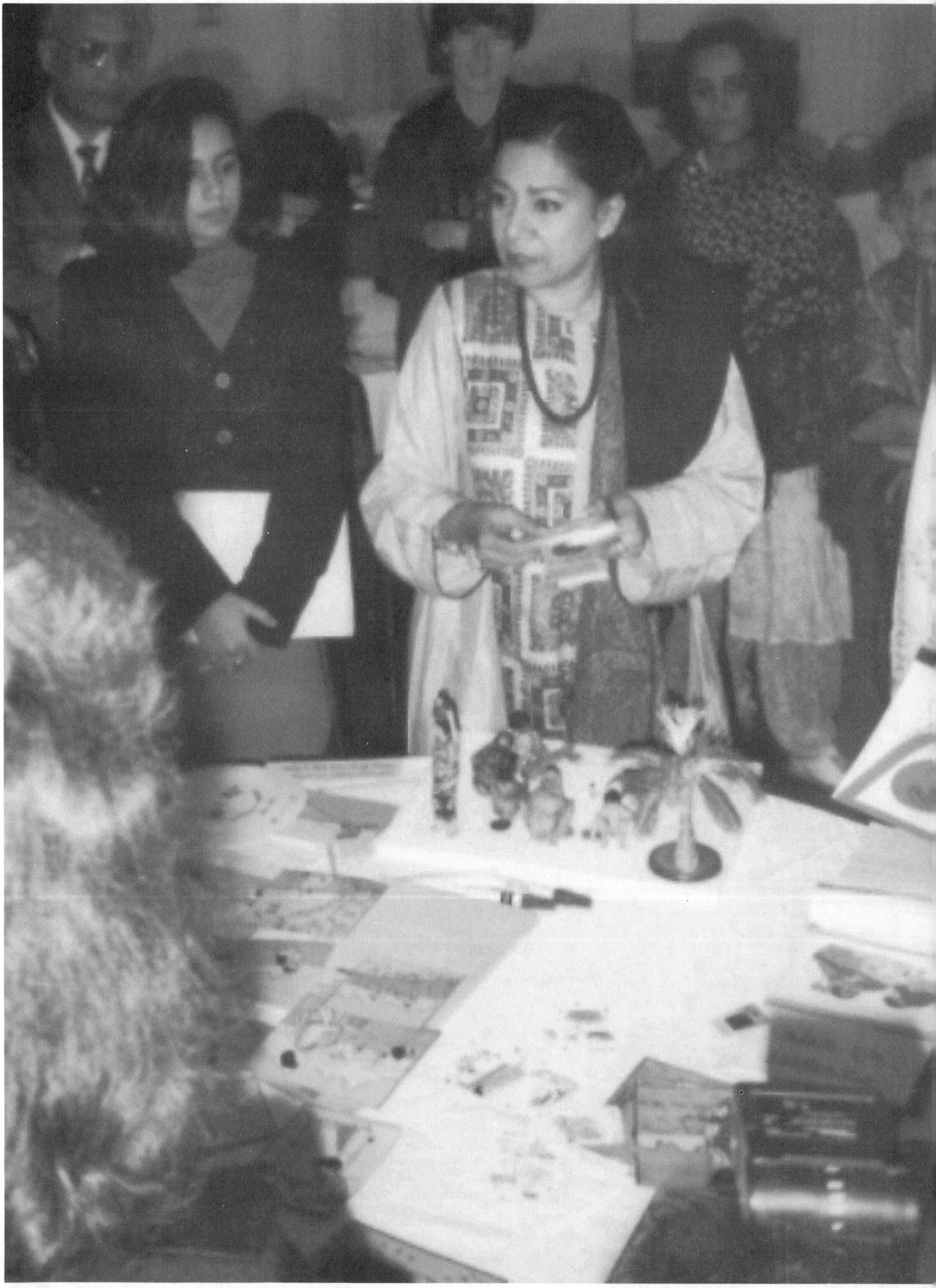

Chapter 5

Choosing Data Collection Methods

No single data collection method is superior to another; rather, each is most appropriate to gathering certain kinds of information. In deciding which data collection methods to use in a study, the logical starting point is defining the kinds of information needed and then choosing data collection techniques best suited to generating that information.

There is often acrimonious debate between those who value qualitative, in-depth and descriptive data and those who want quantifiable "hard" facts. Often, this is a false dichotomy; in most situations, it is useful to obtain both qualitative and quantitative information.

When conducting participatory research, the issue of qualitative and quantitative data is best addressed by directly focusing on the type of information needed. In quick, short-cut studies, using a mix of techniques is usually more appropriate than relying solely on a single data collection method.

In determining what data collection techniques to use, and the degree to which they should be used in a conventional, short cut or participatory manner, there are several issues to address. These include:

- How sensitive the information is within the local cultural context;

- The degree to which the techniques is or can be obtrusive or unobtrusive, direct or indirect, closed or open ended;

- Matching the method with the specific information needed.

What is sensitive information?

A key question in selecting the appropriate data collection techniques is whether the information sought is "sensitive." Sensitive information is broadly defined as information that cannot be obtained directly because people's responses to questions are influenced by factors or motives which result in unreliable or inaccurate data (what economists call "strategic bias"). As a result, respondents may skew their answers to questions in ways that undermine the reliability of data.

> **Chapter Contents**
>
> *This chapter explores the following issues:*
>
> - What is sensitive information?
>   - Desire to please or to appear knowledgeable
>   - Memory lapse
>   - Desire to misinform
>   - Desire to give "correct" answers
>   - Hope to be a recipient
>   - People do not know
> - Characteristics of data collection methods
>   - Obtrusive or unobtrusive
>   - Direct or indirect
>   - Closed or open-ended
> - Matching methods with information needed
>   - Observation
>   - Interviews
>   - Workshops and discussion groups
>   - Semi-projective techniques
>   - Engineering techniques

At the planning stage, therefore, it is important to rate the sensitivity of information in order to ensure that the results of the inquiry will be reliable and valid. This will also permit some short cuts to be taken, since quicker, direct methods can be used when information is not sensitive. More indirect, unobtrusive methods — which may be more time consuming — will have to be used when information is sensitive.

Information that may be regarded as sensitive in one region or cultural context may not be sensitive in another. For example, in some countries in West Africa and in Nepal rural people do not like to have their children counted. In these environments, even an apparently simple question like "How many children do you have?" produces false answers. In other countries, however, parents willingly provide details about their children. Many issues related to sanitation are private, and hence sensitive. The cultural concepts of 'purity and pollution' which influence siting of water and sanitation facilities, and personal cleanliness mores, are particularly important.

The quickest way of determining whether the information that will be collected within the study is sensitive is by asking local people. For example, in a study in rural Botswana which used a mix of participatory and indirect techniques, local extension workers said that any questions related to "wealth" would be very sensitive. Hence, all direct questions about cattle and land holdings were dropped. The issue of wealth was explored using incomplete sentences that people completed by saying the first thing that came to their minds.

Field Insight: Treating Sensitive Information Sensitively, Zambia

Researchers in the Zambia Participatory Poverty Assessment learned that no matter how much pre-planning has been done, flexibility in utilization of methods must be retained to understand sensitive issues. In one community, for example, women felt uncomfortable talking about other people's wealth or poverty, because it was too sensitive. Instead, after completing a social map of their village, the women agreed to discuss their personal situation if the other women were allowed to only agree or disagree with their assessment. The most striking feature to emerge from the wealth ranking exercises was related to the issue of gender. For example, when wealth ranking was carried out with the entire community, women who were household heads were always in the bottom categories. Women were judged poorest for a variety of reasons: the lack of adult children to provide support; widowhood, divorce or lack of a partner; and age.

Source: *A. Norton, D. Owen, and J.T. Milimo. 1994. Zambia Poverty Assessment. Volume 4: Participatory Poverty Assessment. Draft report. The World Bank: Washington, DC.*

Experience shows that when communities have been involved in the entire research process including the selection of questions and questioners, then the issue of sensitivity of information is not as critical. (Issues concerning the degree of sensitivity of information and appropriate research methods can be found in the box on page 79.)

Desire to please or to appear knowledgeable

When people want to be polite, not give offense or appear knowledgeable, they may give socially desirable answers. Questions which may produce skewed answers include:

Community:

- Do you use the community latrine?
  - Is the health worker good?
  - Are springs used for household purposes?
  - Was the training received by the village caretaker good?
  - Are you involved in the village council?
  - Do you use the new hand pump?
  - If we started a water project, would you participate?

Agency:

- Is your manager effective?
- Do you like working for this agency?
- Are you available to your staff?

Memory lapse

Some information may not be recalled because it is not of central importance in daily life. This might include, for example, data responsive to questions such as, "Did your child have diarrhoea in the last two weeks?", "What water sources do you use in the rainy season?" and "When was the last evaluation conducted?"

Desire to misinform

People may not give accurate information because they may consider it is in their interests to obscure the truth. For a variety of reasons, people may not want to accurately answer questions such as: "How many cattle do you have?" "Do you have your baby weighed at the health post?" "Does the caretaker hold monthly meetings?" "Do you pay your fees regularly?" "Is everyone allowed to use this standpost?" "What is the financial performance of the agency?" "How are contracts awarded to construction firms?"

How Information Sensitivity Can Produce Misleading Data

Information is sensitive if it cannot be correctly obtained by outsiders asking direct questions because people:

- Desire to please or appear knowledgeable
- May not recall (memory lapse)
- Desire to misinform
- Desire to give "correct" answers
- Hope to be a recipient
- Do not know the information

What Information is Needed?

Each study requires that certain kinds of information are gathered; determining which types of information are necessary is a key to success. Below are some broad categories of information needed. A single study will not usually require that all of these categories be researched.

Within communities

- Demand-felt need for improved water, sanitation or health
- Village setting
- Infrastructure
- Demographic factors
- Economic factors
- Ecological situation
- Social factors
- Health
- Village level organizations (formal and informal)
- History of community participation
- Priority "felt needs"
- Traditional roles of men, women and children
- Available technology and resources
- Education and exposure to media
- Existing water sources
- Environmental sanitation
- Health and hygiene practices

Within agencies

- Goals, mandate
- Autonomy
- Capacity; skills, resources
- Incentives; organizational culture, orientation
- Organizational structure
- Skills (internal)
- Financial flow
- Field presence
- User-orientation
- Gender awareness
- Ability to be responsive
- Performance criteria; personnel, agency

Desire to give "correct" answers

Respondents may want to be seen as giving the "right" answer to a question, even if it does not accurately reflect their situation. For example, in one district an active health education campaign had been mounted to urge people to wash their hands after defecation and to boil their water. In this context, asking people directly if they practice handwashing is likely to elicit an affirmative response because people know that they are supposed to do so.

Asking trained caretakers who have been supplied tools whether they regularly tighten and oil nuts and bolts in a hand pump may elicit a high percent of false positives (answers of "yes"). This may be the case even in a situation in which caretakers lack the tools necessary to do so.

At the agency level, staff and management may talk with great eloquence about the central importance of involving community people and of the demand-orientation, without any intention to change any practices.

Hope to be a recipient

In many situations, people may feel that if they answer questions in certain ways they will be able to obtain assistance. For example, it is not difficult to imagine how people might answer the following questions in ways most likely to induce the possibility of assistance: "Is water a problem here?" "Is there need for a health post here?" "Do people have sufficient seeds to grow vegetables?" "Will you pay toward the construction of a new water system?"

People do not know

There are some types of information which people may simply not know. For example: "How many minutes does it take to fetch water?" "How many liters of waters do you use every day?" "What are the performance evaluation criteria for extension workers?"

Characteristics of data collection methods

Most data collection techniques are neither inherently conventional nor participatory. Rather, it is how they are used that determines whether they are or are not participatory. For example, mapping of water sources can be conducted by outsiders or by community people.

In determining how best to collect the data, collection methods should be judged on the degree to which they are:

- Obtrusive or unobtrusive

- Direct or indirect

- Closed or open-ended.

Obtrusive or unobtrusive

Obtrusive measures are those which interfere with normal activities and thus will be noticed by the respondents. As a result, these techniques may cause a reaction which alters the usual behavior or situation.

An observer who sits outside a community latrine writing down information every time somebody enters is an obtrusive presence. Because of the observer, some people may avoid coming to the latrine because they feel shy; others who do not usually use the latrine may come out of curiosity.

The same observer's presence could be made unobtrusive if he or she watches casually from a distance; alternatively, the observer could remain posted near the latrine over a number of days, but begin recording data only after a few days have passed and people have become accustomed to his or her presence. A less obtrusive method is when the monitoring is done by a same sex observer from the community.

Another unobtrusive method for monitoring latrine use could be observing the presence of flies or water (if water is used for cleans-

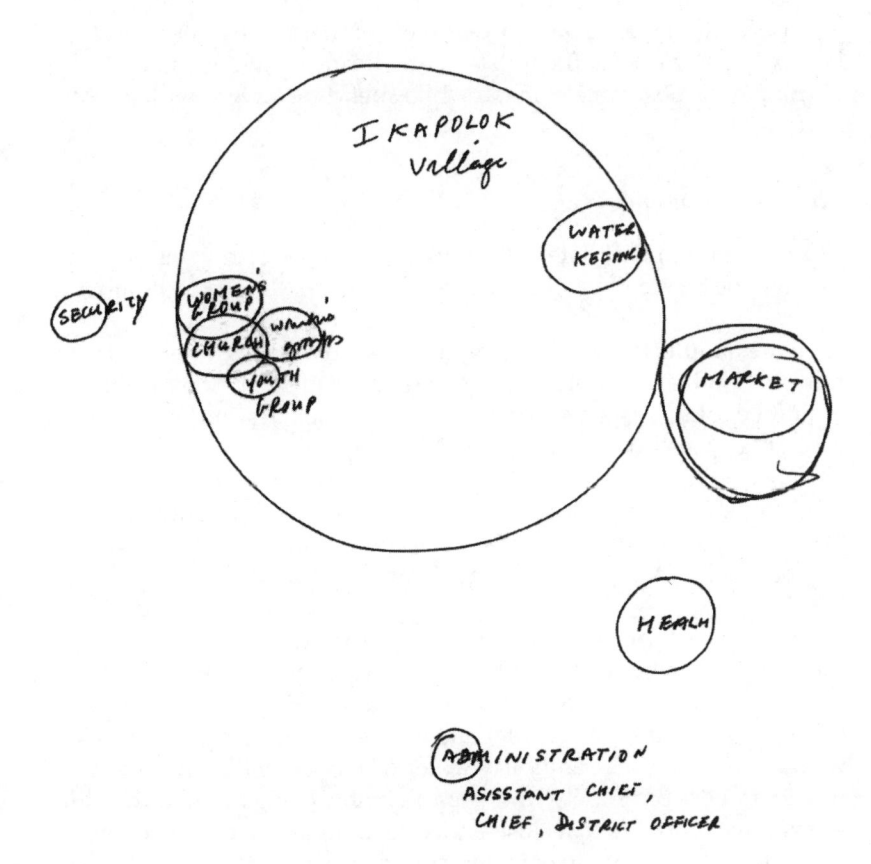

From Draft Field Report, Busia District, KENYA PPA, 1994.

Field Insight: Institutional Analysis by Community Groups, Kenya

Institutional analysis can be conducted by those within an organization or by users. Venn diagrams, or "chapati diagrams," consist of drawing institutions as circles. By mapping institutions, their inter-connection and distance from users, users at the community level can share their perspective of local institutions and the roles they play in their lives. Overlapping circles, for example, depict interaction and shared membership between institutions.

In Ikapolok village in rural Kenya, the chapati diagram activity revealed very detailed information about institutions considered absent or ineffective by the village. Villagers identified the water user group at the borehole, for example, as ineffective because the water did not serve the entire village and the water dried up in the summer. The health center was located 5 kms. outside the village and was not rated satisfactory because it was frequently out of drugs. People also cited fees as a problem for inpatients. Although there was no church building, church groups held meetings and prayers in people's homes. Both local denominations had a woman's group active in village self-help activities.

The village community perceived the district administration, the district officers, the chief and the assistant chief as too distant from them. People said it usually took more than a week to contact any of them with a problem. The other major problem cited was the distance to the nearest police post, which was 7 kms. away. While participants said they wished they could have better police protection, they also cited harassment by the police as a major problem. Apparently, under the guise of stopping smuggling across the Kenya-Uganda border, police often confiscated local property, especially livestock.

Source: *condensed from Field Report, Ikapolok village, Kenya Participatory Poverty Assessment, Kenya, World Bank, 1994.*

> **Matching Methods with Information Needed**
>
> In determining which techniques of data collection are best utilized in a given situation, it is important to first assess the following:
>
> - Purpose of the study
> - Types of information needed
> - Priority of information needed
> - Desired degree of accuracy and detail
> - Information sensitivity
> - The degree to which the methods can be obtrusive, direct, close-ended.
>
> Having assessed these factors, data collection methods should be chosen and structured to achieve the best fit to the kinds of information needed.

ing). When a toilet is not used, it is usually dry and will have no smell or flies and will look abandoned, with overgrown grass in pathways.

Direct or indirect

A direct data collection technique is one which focuses specifically on the issue of interest. Using indirect techniques makes it more difficult for people to give socially desirable answers or deliberately incorrect answers because they do not know the real purpose of the research.

Following are two examples illustrating the distinction between direct and indirect methods.

1. A study is designed to determine whether a women's organization, called BDP, is perceived by community members as useful or important. A direct approach would be to ask, "Is BDP an important organization in this community?" An indirect question would be, "What are the important groups in this community?"

2. In assessing water use in a toilet, a very direct approach would be to ask, "Do you use water for cleansing after defecation?" A less direct method would be to ask, "What material do you use for cleansing after defecation?" An indirect approach to gaining the information would be to observe water use in the household to see if water use for the toilet occurs during a day. Another method is observation made while spending a night with a local family.

Closed or open-ended

Closed techniques of data collection limit the possible types of data that will be elicited to a greater extent than open-ended techniques.

For example, in a closed approach to study the use of oral rehydration therapy (ORT) for diarrhea management, the respondent would be asked to answer "yes" or "no" to the question, "Do you give ORT when your child has diarrhea?"

An open-ended technique would employ questions such as, "What do you do when your child has diarrhea?" and "Do you give any particular food, liquids or medicines?"

A story or puppet play could be used to elicit an understanding of the relationship between drinking dirty water and diarrhea. For example, a play might tell the story of two female puppets and a sick child puppet, with the older female telling the mother that the child was sick because he drank dirty water. An open-ended approach would end the program with the participants being asked if the older woman was correct. In a close-ended technique, the organizer would give a lecture on why dirty water causes disease without much discussion from the participants.

In general, obtrusive, direct and closed approaches are often easier, less time-consuming and less expensive methods of data collection. However, these techniques are inappropriate when applied to eliciting sensitive information in which a high degree of accuracy and detail is required.

Matching methods with information needed

Five methods of data collection, some more commonly used than others, are:

- Observation
- Interviews
- Workshops and discussion groups
- Semi-projective techniques
- Engineering techniques

Observation

As the word implies, observation is simply seeing and recording in some fashion what is spontaneously happening in a particular situation.

Simple as it sounds, observation is probably the single most important technique for collecting data when the information is considered sensitive and when a high degree of reliability and accuracy is desired. Observational methods can be used to gather information about physical conditions and behavioral practices.

Field Insight: Participant Observation Facilitates Institutional Reform, Pakistan

Many World Bank studies have recognized the existence of informal organizations "below the surface of client institutions." In the Second Karachi Water Supply and Sanitation Project, however, the Bank went beyond the initial step of recognition to using a participant observer to understand the elusive and complex characteristics of institutions and to implement a process of change.

Both the government and the World Bank agreed that a large urban water utility was ineffective and in need of institutional reform. The utility was characterized by an emphasis on personal agendas, selective sharing of information, interpersonal and intergroup rivalries and power struggles, intimidation, and coercion. Since these and other problems were fundamentally behavioral and attitudinal in nature, any institutional changes would have to be initiated at a behavioral level through involving local staff in planning and implementation. A participant observer was placed in the utility. The participant observer's role, therefore, was not only to determine how the organizations functioned, but to serve as a "process consultant" helping the staff to examine and evaluate "what was being done, how it was being done, and by whom, against agreed action plans."

To this end, the participant observer took part in a variety of tasks including attending meetings, developing checklists, monitoring progress, identifying and clarifying problems, and through informal discussions, making sure information was being disseminated, facilitating communication, and building trust. Observation, for example, revealed that 174 clerical steps were required before any payment could be made to a contractor.

Client staff participated fully in the process creating an increased sense of ownership of outcomes and decisions. Over time, defenses eventually broke down and trust developed, with fewer and fewer attempts to "mask reality."

The participation of the local institutions and their staff in planning institutional reform resulted in the adoption of appropriately tailored solutions and improved communication with the Bank. The process of change has already induced new behaviors of transparency and accountability at the individual level, and some similar changes are now being seen at the institutional level.

Source: N. Boyle. 1994, personal communication; and Boyle and Albert Wight 1992. "Policy Reform: The Role of Informal Organizations." World Bank. Infrastructure and Urban Development Department. Urban No. OU-5, September.

Physical conditions that can be observed as basic information for use in planning include: the settlement pattern in a village, soil conditions, topography, presence of rivers, wells, pipes, taps, springs, and sanitation. Observation is essential to establish the functioning and utilization of new water supply and sanitation facilities.

Functioning. Because they do not know or because they want to present a positive picture of facilities, leaders often report facilities to be functioning when they are in fact broken down. Site visits can establish accurately whether a toilet or handpump is in working condition or not. For example, at a village officials meeting, a report was given indicating that a community toilet in rural Timor was heavily used; however, on a site visit, it was found to be broken down and abandoned. When asked about the toilet, villagers laughed and said that the only people who had ever used it were visitors.

At the agency level, as demonstrated by the Pakistan example (above), much can be learned about how agencies initially function before developing strategies for change.

Utilization. Use of facilities can be more accurately estimated by direct observation than through sole reliance on verbal reports.

Behavior patterns and practices — such as hygiene practices, who fetches water, how water is carried, how containers are cleaned, activities and interaction at source — can usually be more easily understood by observation. First-hand observations can also reveal who attends clinics and health posts and why they do so. How decisions are made in existing community users groups or water committees is usually better revealed and understood by observing a meeting than by collecting information through interviews that require recall. Behavior of health personnel toward poor people, or the amount of time it takes and the number of steps involved in paying a water bill, getting an inaccurate bill corrected, or obtaining a receipt, can be very revealing.

There are, however, some important limitations on the effectiveness of observation techniques:

- They can be time-consuming;

- They can result in a focus on the atypical if, for example, a few selected and well-functioning toilets or handpumps are chosen by the village chief for observation studies;

- Observing only villages, facilities or houses that are close to motorable roads can produce "tarmac bias;"

- Selective perception can result in missing the entire picture of what is occurring. While attending a meeting, an observer may focus on the few people who spoke and forget those who did not;

- By definition, observation only deals with what is occurring now, not what has taken place in the past;

- Observation results in descriptive information which has to be interpreted. Assigning cultural meaning to an episode, incident, behavioral patterns or physical conditions can be subject to error. In a village in India, for example, school children were observed to take water to school in bottles which were often coated with green slime on the inside. The observer concluded there was no drinking water in school and children drank from the dirty bottles, but when asked, the children revealed they were required to bring water for the school garden;

- The quality of observation depends on the training of the observer.

Types of observation. Some of the limitations of observation can be overcome by selecting between two main types of observation — participant and structured.

Participant observation occurs when the observer participates to a greater or lesser extent in the situation being observed. This approach is typically used by anthropologists who live in the community being studied for months or even years on end.

In a water and sanitation study, extended community stays are rarely possible. However, staying overnight for just a day or two with families may reveal behavior, practices, values and conditions that could not have been discovered in a short visit. For example, an observer staying with a family for two nights in a south Asian village found hygiene practices such as walking past the outdoor toilet to use cornfields, not washing hands after defecation, cooking food that had only been lightly rinsed, and family members who needed to urinate at night squatting in a corner of the room.

To be successful, participant observers need a high level of self awareness, a good memory, a respect for people and a non-judgmental attitude. Without these, participant observation can result in a distorted picture of a situation because of selective perception, poor memory and inability to attach valid cultural meaning to observed behavior.

Structured observation. Observations can be systematically structured using observational sheets or other methods to note particular types of behavior over a predetermined period of time.

In Bangladesh, non-literate village women observed the number of times particular families collected water and the number of pump strokes used. They did this by transferring pebbles from one can to another for each stroke of a pump.

Structured observations can be utilized to gather a variety of information: who collects water; time taken for a water journey; number of trips made per household; amount of water collected per day; frequency and length of time of use of different water sources or sanitation facilities; decisionmaking or participation in a group meeting; hygiene practices in the household, including hand washing; handling of children's faeces; rating of physical condition of water and sanitation facilities.

Interviews

When two people are in conversation, and one person is asking most of the questions or controlling the situation, it can be labeled an "interview." Interviews, when combined with information obtained through observation, can lead to an understanding of a situation or to correctly interpreting a description of a situation or event.

Interviews vary primarily in the degree of predetermined form and structure that is imposed on them, ranging along a continuum: open-ended, focused, semi-structured and structured.

Open-ended interviews

Open-ended interviews are free-flowing to a greater extent than other types of interviews. The interviewer has a topic in mind and the interviewee is encouraged to talk about the topic. The interview is conducted in a very conversational, informal manner. The advantage of this method is that the lack of predetermined structure allows free association resulting in a broad exploration of a topic which may bring to light unknown issues and facts.

The most important interviewer qualifications for a good unstructured interview are respect for the interviewee, being a good listener, ability to tolerate silence and ability to probe and ask questions in a stimulating, non-threatening fashion.

For example, in one interview, an informal conversation about water sources and a failed hydropower dam project — including long periods of silence by both the interviewer and interviewee — resulted in the respondent finally talking about "mythical" owners of water who had not been appeased by the project implementors.

In another case, the desire to understand the relationships and roles of traditional medicine, faith healing and modern medicine led a field worker to seek a meeting with a faith healer. A long interview resulted from the statement "I have heard about your powers and have come to learn about them from you." The researcher hardly said anything except tolerating periods of silence, making sounds of encouragement, nods and asking for a few clarifications.

Open-ended interviews are useful for exploring ideas and hypotheses; when a situation is not clear or does not make sense; when an investigator wants to learn about what is important to people in a certain context without suggesting any answers by posing particular questions. In this form, open-ended interviews become "listening surveys".

The limitations of open-ended interviews are that they are time consuming, they may or may not result in "relevant information," and they require greater interviewing skills than more structured interviews.

Focused interviews

As the word suggests, these interviews focus on a particular topic and are carried out with a checklist of subtopics or issues to be covered. Focused interviews have many of the same advantages as open-ended interviews. In addition, the security of a checklist instills confidence in interviewers who may otherwise be afraid of venturing into the unknown. However, because of a relative lack of structure, focused interviews still require fairly skilled interviewers.

A checklist for conducting a focused interview to learn about preferred sanitation facilities and their use could include:

- Different types of groups
- Membership in group
- How and why joined group or did not join group
- Perceived function of group
- Group leaders and their style of leadership
- Person's role in groups - what has the person done or participated in
- Past achievements of group
- Present activities of group
- Future plans
- Problems experienced
- Possible solutions

In a focused interview, the interviewer sets the interview parameters by asking certain pointed questions about a topic. The interviewer then follows the lead of the interviewee, asks additional questions and retains control over the direction of the interview without breaking the flow of conversation. The order of questions is not important. However, by the end of the interview all the topics should be addressed.

A few well-conducted focused interviews with a range of people may reveal more about the functioning of a water group than a more formal house-to-house survey. Notes are often not made during the interview but are recorded afterward.

Semi-structured interviews

Semi-structured interviews are more structured than focused interviews. They generally consist of a series of open-ended questions that are asked in a predetermined order. However, each question is followed with additional probes until the answer is explored in some depth.

Semi-structured interviews also require carefully trained interviewers. However, since the focus is on questions in an interview schedule, it is easier to train relatively inexperienced people in a short time to do quality interviews. Even though a predetermined format of questions is followed, the interview is conducted in a

> **Field Insight: Use of Key Informants, Indonesia**
>
> A number of key informant interviews were done with a range of people, both leaders and ordinary community members, living away from the main road. A conversation with an older man living far away from the road revealed recent government resettlement schemes which encouraged people to live close to roads. The conversation revealed how an entire area of thirty households had been forced to move close to the roads where there was no water source and how twenty-five of these households had moved back to their old settlement to be close to a perennial spring. Until that point, not a single leader or "ordinary" community person had mentioned resettlement schemes.

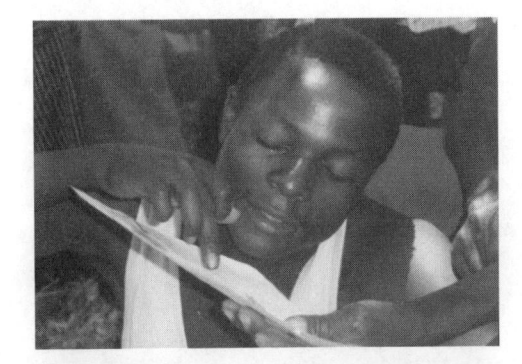

conversational manner. Notes are usually taken during or soon after completion of the interview. The purpose of note taking as an aid to memory should be explained to the person being interviewed. It is also useful to state that no names will be reported and that the information will be aggregated.

In structuring an interview schedule, the following general points should be kept in mind:

- Introductions are very important to help people relax, to establish rapport, to be accepted and to establish credibility. Introductions should contain factual information about the interviewer and the purpose of the interview. Introductions should assure privacy and confidentiality of all responses.

- Interviews should start with facts, questions that are easy to understand and answers that are not controversial.

- Questions should move from general to specific. For example:

    - What is the total number of your children?

    - How many of your children are between the ages of 0-5 years?

    - Who usually collects water?

    - What are the ages of children who collect water?

The questions should include probes to clarify a situation without suggesting answers or drawing conclusions not stated by the interviewee.

Example 1

If the answer to a question is "I make more water trips in the dry season than in the rainy season," an appropriate probe would be, "Why is that?"

An inappropriate follow-up would be, "So you use water that collects in your well that is dry now. For what do you use this well water?"

Example 2

Assume that the response to a question is:

"Nothing goes smoothly in the village because the village head does not care, he sits on the money."

An appropriate probe could be:

Facial expression showing interest, nodding, keeping silent and letting the person elaborate. If needed, ask follow-up questions such as: "Can you explain what you mean?" "Can you give me an example?" "Has something like that happened recently which makes you feel that way?"

Inappropriate responses include showing disapproval or any obvious emotion, positive or negative, such as:

"You are accusing the village head of stealing money," or, "Yes, I understand that the village head is a drunk."

Other general points to keep in mind when structuring interviews include:

- Avoid double negatives such as: "Is it true that you are not dissatisfied with the work of the extension workers?"

- Ask neutral questions. For example: "Now that the construction work is over, how do you feel about the role of the extension worker?"

Leading questions can be asked occasionally to help people start talking about situations or facts that would otherwise not be polite or proper.

Example 1

In a village, there were strong beliefs about "spirit owners" of water sources and the need to obtain permission from spirits through certain rituals prior to construction at water sources. When the issue of rituals was raised directly, people denied their existence because, as nominal Christians, they could not acknowledge these practices. However, assuming the existence of rituals and asking "leading questions" produced long discussions.

A neutral approach would be:

Q: "Before construction, were there any ceremonies that were performed?"

A: "Yes, the elders of the church said prayers."

A leading question could be:

"Before the construction was started, how important was it for the village elders to kill a pig and do a ceremony at the source?"

Example 2

Village "A" was experiencing acute water distribution problems with neighboring village "B". Pipes carried water to both systems, but because village "B" kept taps open all the time, village "A" received hardly any water. The relationship between the two villages was strained and the atmosphere bitter.

Field Insight: Limitations of Structured Interviews, Ghana

In an extensive water use survey in northern Ghana, researchers employed a number of different data gathering techniques, including observation (dusk to dawn surveillance of major water sources); mapping of the survey area (by triangulation); using these maps to randomly select respondents; and conducting structured interviews with women. However, despite the generally high quality of work and the use of a number of quality control procedures, problems still occurred which reflect some of the limitations of the structured interview technique.

Women had difficulty answering many of the questions. For example, many appeared to have genuine problems in replying to questions that asked about the **usual** number of water collecting trips made each day. This ran counter to the women's normal way of thinking about collecting water — that the number of trips made to collect water depended on a range of factors such as need, other activities which must be completed in the day, storage capacity, visitors, a woman's strength and so forth. For some women, it appeared there simply was no such thing as a "usual" number of trips.

Researchers also encountered problems with questions that investigated such areas as bathing and children's health. Responding to questions regarding bathing, women replied that they used large amounts of water. However, social norms led respondents to exaggerate the number of baths they and their family took. In addition, asking questions about children's health created an expectation among respondents that researchers had medicines to dispense. This expectation in turn distorted respondents replies.

Source: Malone Given Parsons Ltd. *Ghana Upper Region Water Programme Evaluation Project. Report 5: Technical Appendix One, Survey Methodology.* Prepared for Canadian International Development Agency.

Conducting Semi-Structured Interviews - Two Examples

Example 1

Purpose of interview:

To study preferred sanitation facilities and their uses.

Interview question:

Q1. Different people in this community prefer to defecate in certain places or use certain defecation facilities. Which are the most common facilities in this community?

Possible answer:

Some people go near the railroad, some behind the coconut grove and some have toilets outside the compound.

Problem:

Q: Are there any other facilities used?

A: Yes, I think one public toilet was built on the other side of the village.

Q2. Different people prefer different facilities. I want to talk to you about each type of facility individually because each facility has its own advantages and disadvantages. Let us start with the railroad.

Q3. What are the advantages or good things about using the area near the railroad for defecation?

Q4. What are the problems or disadvantages about using the area near the railroad?

Q5. How about the coconut grove? What are the advantages or good things about using the coconut grove for defecation?

Example 2

Purpose of interview:

To assess involvement in water users groups.

Interview schedule:

Q1. Are there any water users groups in this village? Yes No

Q2. a. How does someone become a member of the group?

  b. Can anyone become a member of the group?

Q3. Do you belong to a water user group?

Q4. Why do you belong to a group? Why did you join the group?

Q5. Why was the group formed? What is the purpose of the water groups?

Q6. Has the group had any successes? Has it achieved any results? Why has the group experienced problems?

Q7. Why has the group been successful?

Q8. Who selected the leaders of the group? How were the group leaders selected?

Q9. How were decisions made in the group about:

- monthly contributions
- sanctions
- water distribution?

Neutral approach:

Q: "Are there any problems with the village below about distribution of water?"

A: "Oh, no we share the water."

Leading question:

Q: How and when did the problem with village "B" begin?

A: "Mrs. Benny sent her son and he came quietly and broke the lock on our controlling valve."

Structured interviews

A structured interview that allows little room for asking additional questions or probes is at the opposite end of the continuum from a free-flowing or open-ended interview.

Structured interviews or questionnaires can be useful when:

- Much descriptive information is already available that enables correct interpretation to answers of close-ended questions.

- The existing data base is substantial, and allows a clear delineation of appropriate questions to be asked.

- The primary interest is in quick quantification of narrowly defined topics.

- People can be expected to answer questions frankly and honestly and to know the information.

A questionnaire should be formulated with care, keeping in mind the same issues as in developing an interview schedule. General guidelines for questionnaire development include:

- Keep questionnaires short.

- Each question should express one idea only.

- Avoid double negatives.

- Ask neutral questions.

- Start from general questions, move to specific questions.

- If there are open-ended questions, there should be sufficient space on the questionnaire to note the answers.

- Categories of answers should be exhaustive and mutually exclusive so that decisions on how to categorize a response are clear cut.

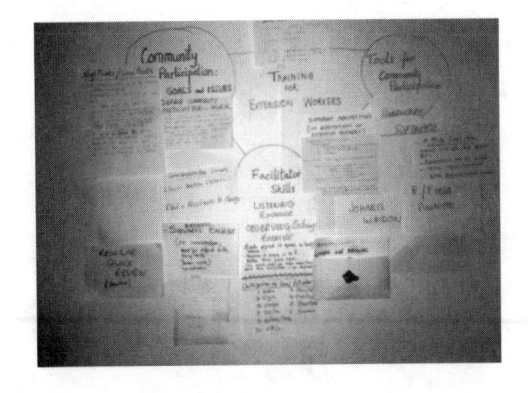

Field Insight: Managing a Group Process

Group facilitation is an important skill for participatory trainers and researchers. In large workshops, most work is usually done in small groups. The following strategies used with sensitivity to cultural differences are effective in managing problems of domination of groups by a few people, drawing the more reserved into the group process, and keeping interest in the group process alive.

- Make participants aware from the start that the formation of small groups has a dual purpose: to draw upon the experiences of every member of the group; and to give everyone an opportunity to participate.
- If group interaction is floundering or being derailed, ask participants to brainstorm, and formulate their own ground rules for interaction. Ask them to post the ground rules as a reminder to monitor their own behavior.
- Assign tasks that require the group to work in still smaller groups. This automatically breaks up dominant behavior.
- Reorganize the composition of the small group so that participants get an opportunity to interact with different people.
- Use hands-on materials which provoke thinking, creativity, and the expression of individual viewpoints.

Source: *Lyra Srinivasan, personal communication.*

Workshops and discussion groups

Properly used, workshops and discussion groups can provide valuable information at any stage in a program cycle. For example:

Planning. In one country, a workshop brought together senior officials from relevant ministries prior to a study. The purpose was to identify: (1) issues or indicators of interest to planners; (2) existing information; (3) resources (especially personnel) available from different departments; and (4) to discuss research methodologies, especially research sites. An important unstated objective was to generate interest in and build commitment to the study, thus increasing the likelihood that results would be utilized.

Before training. A workshop was held in India to identify problems and training needs of drillers and to develop a training plan for drilling teams. The workshop brought together experienced drillers and trainees for a day.

Evaluation. A workshop involving school principals, teachers, parents and students was held in Lesotho to discuss the school latrine program. By emphasizing dialogue among the participants, the workshop was able to develop an evaluation strategy for the school latrine program.

The number of participants in a discussion group can range from two to sixty. However, when there are large numbers of people it is essential to organize them into several smaller groups to provide opportunities for greater informal interaction.

The role of the leader or facilitator in discussion groups and workshops is crucial. When the primary purpose of such groups is for data collection or gauging people's preferences, ideas, assessments and commitments, it is essential that the leaders act as facilitators, rather than dominate the proceedings. The role of the facilitator is to structure the group in such a way that people talk freely, exchange ideas and share them with the group.

Facilitators should pay careful attention to the following factors to ensure that they are used advantageously rather than allowed to become disadvantages:

Size of groups. Large numbers of people can generate energy, excitement and commitment. However this energy has to be channeled. This can be done by creating smaller groups and giving everyone clearly defined tasks; for example, to draw a picture of their local water source or describe in writing the five factors that most affect their job performance negatively. When large numbers of people participate, it is important to have more than one facilitator.

Mutual respect. Perhaps the factor that most inhibits people from speaking is when they feel they may be wrong and will be laughed at or ridiculed. If sharing of information or constructive interaction is expected, it is important to emphasize a non-evaluative attitude at the individual level. Depending on the nature of the group and purpose of the workshop, this can be created by warm-up introductory exercises emphasizing cooperation and different perceptions of the same object.

Gender. The gender composition of a group — both participants and the facilitator — can be very important depending on the cultural context and the purpose of the meeting. A female facilitator may be more effective with a group composed primarily of women. In mixed-gender settings, it may be important to form separate groups of men and women to ensure that women actively participate, especially during the early stages of the meeting.

Status. Many groups become less than optimally effective because status differences between participants limit the free exchange and discussion of ideas and information. For example, in a planning meeting in Nepal, village level workers dominated village women until the two groups were separated. The same may be true at workshops held at ministerial levels involving officials of different ranks.

Cultural norms. In some settings, cultural norms may dictate that one or two group members be spokespeople, or that contradicting what someone else has said is inappropriate. In these cases, facilitators must set the stage with appropriate introductions and by utilizing activities that do not force people to contradict each other in culturally unacceptable ways.

In a village setting, it was found that when wives of village leaders were in attendance, other women in the group became withdrawn. In one situation, the problem was resolved by extensive consultation with the wives prior to the meeting. The wives of leaders then agreed that since much information had already been collected from them, their presence at the group meeting might inhibit others. In another situation, the wife of the village head was assigned the role of a facilitator of the large group. In a third group, the problem was resolved by requesting the leaders to state their opinions and impressions after everyone else in the group had already done so.

Nature of topic. Not every topic relevant to water supply and sanitation can be best understood by holding a discussion group or workshop. Topics that are extremely controversial or considered private are not appropriate for workshops unless the facilitators are confident of generating positive solutions rather than accentuating further divisiveness in a group. Controversial or private topics can also be introduced when a participatory approach is used throughout and the workshop is organized by participants themselves.

Controversial topics may include: water rights; water distribution problems; evaluation of individual performance; and group and village leaders.

Field Insight: Combining Techniques to Develop a Community Payment Schedule, Ghana

In a project in Kumasi, Ghana, the World Bank financed a combination of conventional survey techniques and participatory techniques to come up with an effective payment schedule for community members.

Initially, a household survey on "willingness to pay" was conducted. However, although the research was carried out within the same town in which the project was being implemented, the survey had not, in fact, covered the people in the specific project area. Thus, project stakeholders had not been surveyed.

To correct this information gap, project staff held a series of community meetings to assess how much users would be willing to pay for the project, and how they would do so. The meetings revealed that the most important concern for community members was the size of their **monthly** payments. Project staff originally suggested a two-year payment schedule, but community members felt monthly payments on this schedule were too high. In order to lower the monthly payments to an acceptable level they had to be stretched out over a five year period. This, in turn, raised the total cost burden to the community because of higher interest payments. However, community members were less concerned with this, than the fact that their monthly payments would now be an acceptable amount.

Interestingly, the amount of the monthly payment finally arrived at through the community meetings used in this participatory process was the same as that arrived at through the willingness to pay surveys conducted in the larger town area.

Source: *Albert Wright, TWUWU, personal communication, 1994.*

> **Field Insight: Indigenous Groups Monitor Their Water Quality, Split Lake Cree First Nation and Chol Chol, Chile.**
>
> Following a river diversion in 1970 by Manitoba Hydro, the level of pollution in the Nelson River became so high over a twenty year period that it changed the basic life patterns of the fishing communities along the river and increased the dependence of these remote communities on government assistance. Fishing had to be halted because of high levels of mercury found in fish and drinking water was trucked in; bottled water was used to make infant formula.
>
> Together with colleagues from Environment Canada, Peter Seidl developed a community-based model for using simple and inexpensive water testing methods to empower communities to manage and monitor the quality of their drinking and recreational water. One of the pilot programs was tested in Split Lake, where a water treatment plant was built. Local people were trained as technicians in simple water quality testing methods which they performed weekly rather than monthly. Results were delivered to households within twenty-four hours rather than one month later by distant laboratories. This immediate feedback from people living within the community, motivated follow up action. When bacteriological levels were high, communities took immediate remedial action by changing water storage, water handling and other behavioral practices. Besides reduced cost of water quality monitoring, another indicator of the success of the community based approach was reduction in detected contamination levels from 95 percent in 1989 to 5 percent in 1993.
>
> With support from IDRC, the program's impact was magnified by making it available to native peoples in Chile. Cree First Nation technicians from Canada were flown to Chile to share their experiences with and train Mapuches people in Temuco, Chile, to implement their own community-based water quality monitoring process.
>
> Source: *Peter Seidl, personal communication; Internal World Bank memo, January 5, 1994; and Peter Seidl and B.J. Dutka, 1993.*

Aids to discussion groups and workshops. Facilitators should be well prepared and have a clear idea of the aims, goals and processes involved in the workshop. Selected discussion aids — materials for activities including paper, pencils, answer sheets, typed copies of questions, props for role plays, puppets, and voting games — should be prepared and tested prior to the workshop.

Semi-projective techniques

Semi-projective techniques can be used as aids to understanding either in groups or individually. They are commonly used by researchers in psychology and in anthropology and sociology. In applied development research, semi-projective techniques have been used by practitioners of the participatory approach. Semi-projective techniques provide stimuli on which participants project their own views, perceptions and understanding. Stimuli which are ambiguous or open-ended can be visual materials, incomplete sentences, stories or role plays.

In recent years a variety of games, simulations and other activities have been developed and adapted to a variety of situations. These include puppets, role plays, card and board games, story, picture, or sentence completion, activities involving voting, making of pictures, and posters. These techniques are described in detail in Chapter 6.

Engineering techniques

Much information is already available about specialized techniques such as ground water surveys, water quality testing for total or faecal coliform using portable water testing kits. Even water quality monitoring can be conducted by village groups or students trained to carry out tests.

How Sensitive is the Information?

In developing a data collection strategy, the researcher must determine the degree of sensitivity of the information and consider the qualities of methods needed (can the technique be obtrusive, direct, closed?) The following matrix identifies the key issues associated with degrees of sensitivity of information.

	Socially Sensitive	Potentially Sensitive	Not Sensitive
Type of information	Access to & use of facilities; Functioning of local institutions; Needs, interests, preferences; Leadership (formal/informal); Hygiene behavior & practices; Tradition & extent of community involvement in decisions	Number of wards; Disadvantaged groups; Types of water sources; Number & types of facilities; Functioning of facilities	Number & type of facilities; Community socioeconomic and demographic information; Infrstructure development; Structure of existing institutions
Characterisitcs of data collection method	Indirect Unobtrusive; Closed/open-ended	Indirect/direct Obtrusive/unobtrusive; Closed/open-ended	Direct Obtrusive; Close-ended
Techniques	Observation (structured & participant); Interviews (open-ended, conversational); Semi-projective tools; Workshops, discussion groups	Interviews; Questionnaires	Questionnaires; Official meetings; Site visits; Structured observations
Who collects data	Community members; Field workers; Key people	Investigators; Experts; Key People	Experts; Key people

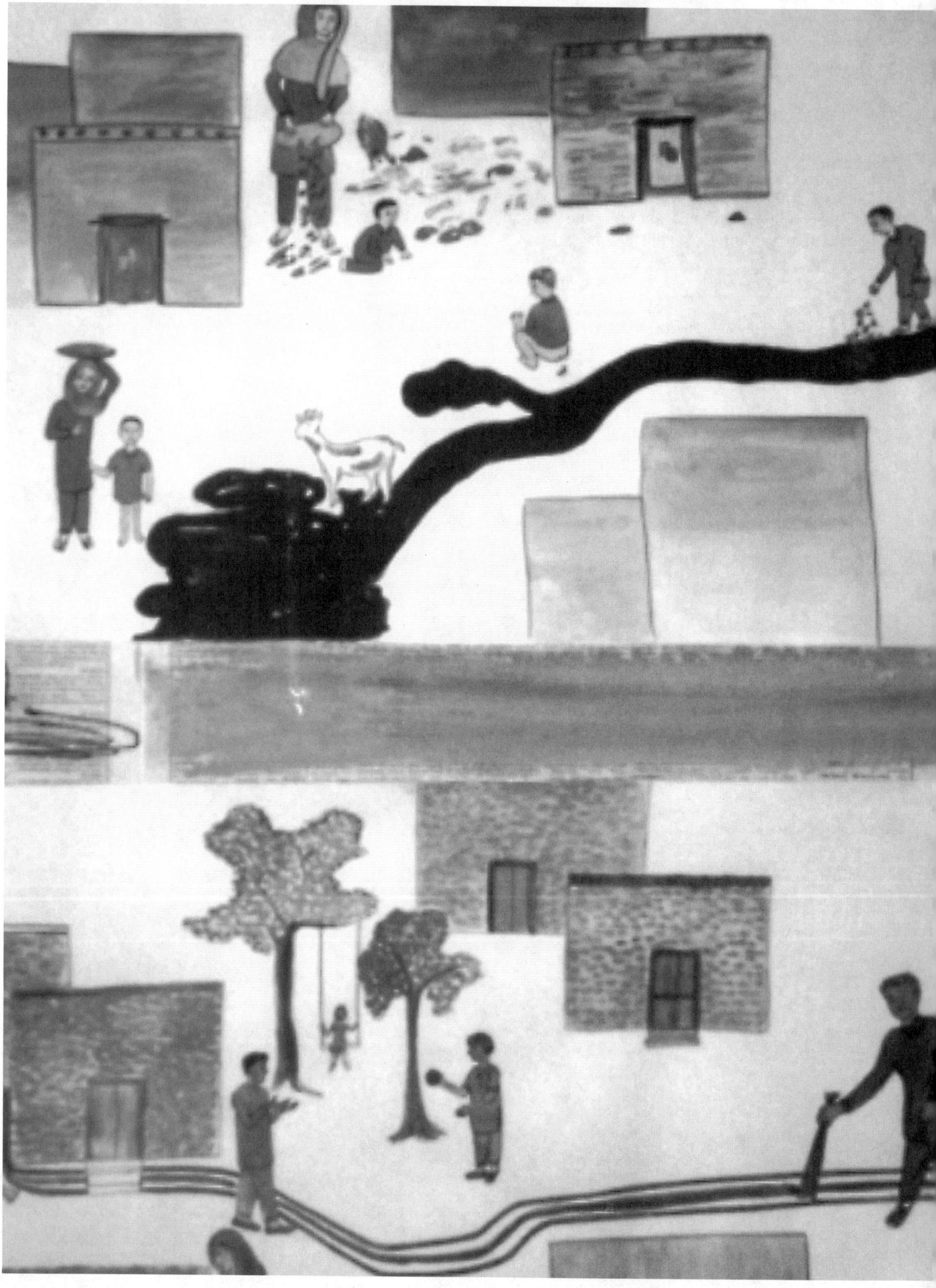

Chapter 6

From the Field: Innovative Data Collection Methods

No researcher can or should use all of these methods in the same setting, at the same time. The methods should be adapted to the particular setting and used as ideas for further innovation, experimentation, and creation. Detailed descriptions of many of these activities and how to use them can be found in the second section of this document.

A word of caution: it is essential that these methods be used as part of a long-term community or staff empowerment process, rather than to extract information from people primarily for external planning or research purposes only. One of the real challenges in adopting participatory approaches is giving up control to participants, facilitating their growth, and assisting local people in solving their own problems. If the materials are used in extractive ways or to preach particular messages, then their application is no longer participatory.

Small-scale models

Use of life-size models of water and sanitation facilities is common in pilot and demonstration activities during implementation of projects. However, small-scale models can also play important roles in the research phase and in the design, construction, monitoring, and evaluation of projects.

In the Maldives, two small models with interchangeable, moveable parts made from locally available materials were used to collect data on user satisfaction with existing community latrines and to facilitate user involvement in designing improvements. One model represented the existing toilet; the other depicted an alternative design. Pieces of plastic, paper, and cardboard were used to try different combinations of roofing, doors, windows, and seat. Presented at group meetings with men and women, the model generated great interest and discussion. People played with the pieces, added others on their own and altered designs until a consensus emerged as to the preferred toilet type. For the researcher, making recommendations about design of improved toilets became a simple task of reporting what people wanted and why.

In Timor in Indonesia, a World Health Organization engineer used home-made, simple wood models of spring captures with moveable parts to involve women in designing and evaluating

> **Chapter Contents**
>
> *This chapter explores the following issues:*
>
> - Innovative research methods developed in the field in Asia, the Middle East, Latin America, and Africa:
>   - Small-scale models
>   - Listening surveys
>   - Walking surveys
>   - Mapping
>   - Stories and folk tales
>   - Indigenous knowledge systems
>   - Pictures and drawings
>   - Photographs
>   - Games and simulations

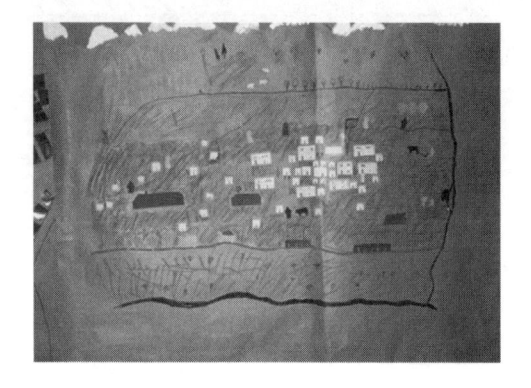

drawbacks of existing spring captures. Skepticism about the ability of the women to design spring captures and related bathing and washing facilities quickly disappeared. Soon the women were spontaneously discussing such complex questions as whether a cement bathing cubicle was worth the additional cost for the extra cement it required. A local technician who observed this meeting went on to a second village to facilitate similar detailed discussions with men and women, this time using even simpler materials such as sticks, mud, paper, leaves, and grass.

In Pakistan, an engineer in the UNDP-World Bank Water and Sanitation Program developed simple wood models to represent different elements of piped water systems to evaluate the functioning of existing systems.

Listening surveys

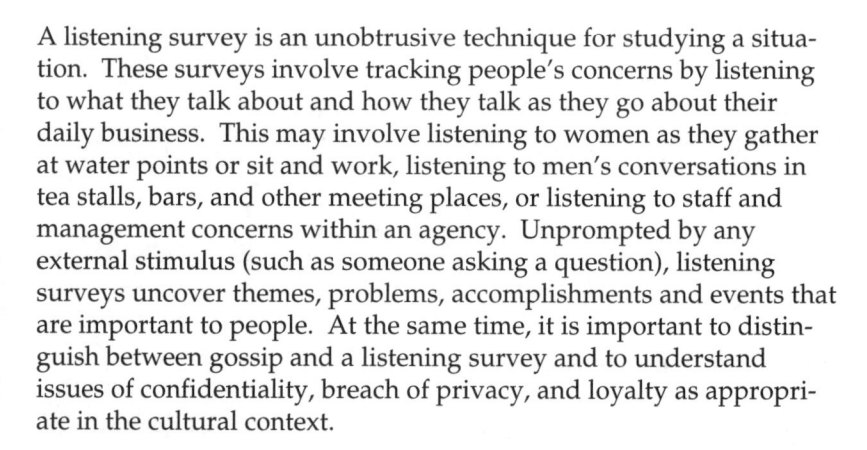

A listening survey is an unobtrusive technique for studying a situation. These surveys involve tracking people's concerns by listening to what they talk about and how they talk as they go about their daily business. This may involve listening to women as they gather at water points or sit and work, listening to men's conversations in tea stalls, bars, and other meeting places, or listening to staff and management concerns within an agency. Unprompted by any external stimulus (such as someone asking a question), listening surveys uncover themes, problems, accomplishments and events that are important to people. At the same time, it is important to distinguish between gossip and a listening survey and to understand issues of confidentiality, breach of privacy, and loyalty as appropriate in the cultural context.

Listening surveys can be conducted by outsiders or insiders. Outsiders should participate in people's lives as naturally as possible; for instance, rather than carry a notebook, the listener should wait until after the interview to write notes to aid memory and analysis. If the listening survey is conducted by community people, there is less chance of disruption. Illiterate village people often have astonishing memories because they have not become dependent on written words for transmitting information or knowledge. If needed, simple devices can be developed to assist the villagers in remembering — drawing symbols to represent themes, and counting the times a theme occurs by piling stones, rice, or corn in front of its symbol.

The findings of the listening survey should be presented to and discussed by the concerned community. In rural Zimbabwe, for example, a needs assessment was carried out through a listening survey which sought to evaluate how the activities of an integrated rural development project fit in with people's priority needs. After a week of listening, the themes that emerged were discussed in a community seminar. This led to an evaluation of project activities and their centrality in people's lives. Other ways of feeding back to the community the themes emerging from a listening survey include

Field Insight: Model Piped Water Systems as an Evaluation Tool, Pakistan

In an exploratory study of World Bank projects in Azad Jammu and Kashmir and Punjab, Pakistani researchers used a variety of participatory tools to understand the functioning of piped water systems under two institutional arrangements. The study was conducted in sixty-nine communities in two provinces and used informal sampling techniques.

In order to understand the history of piped systems, the interaction between the communities and the agencies (the Public Health Engineering Department in Punjab and Local Government in Azad Jammu and Kashmir), a "model" activity was developed by sociologists and engineers working together. The model was made up of several individual parts including houses, wells, handpumps, water tanks, school buildings, dispensary/hospital, water pipes (represented by straws), valves, and standposts.

Researchers held separate data gathering sessions with women and men. Women particularly seemed to be empowered by participating in the activity. In one community, after the women had constructed a model of the village and discussed the issues raised by the researcher's questions, the women stated that this was the first time someone had elicited their opinion on community matters. In addition, they said that the next time a water project or other development project was initiated in the village they would know what questions to ask.

In the male groups, the model helped create an informal and creative environment. Men could easily squat on the ground, argue, and express their views without being intimidated by the presence of an outsider. The model put the community members and the researcher on an equal footing. The model also proved useful in resolving conflicts within the group and ensuring that no one community group dominated the process of information sharing.

Constructing a model proved to be an excellent tool for gathering gender-specific information and was most effective when these simple guidelines were followed: 1) do not introduce the model as a game because participants feel this trivializes the process; 2) "warm up the group" by asking a few open-ended questions before introducing the model; and 3) use the model in groups of eight to ten people (larger groups tend to be dominated by one or two people).

Source: *K. Javayah, K. Minnatullah and R. Zafar, personal communication, 1993.*

Field Insight: Tapping the Different Perspectives of Men and Women, Sierra Leone

Ensuring that a community's diversity—its differences in ethnicity, religion, age, gender, and economic status—is represented in research is important in all participatory data gathering. One group cannot represent the perspectives or knowledge of all the groups within a given community. When field workers call village meetings only with male elders or interview only young men, a skewed perspective of the needs of the community often results.

In Sierra Leone, a mapping exercise conducted with villagers not only uncovered the different felt needs and perceptions of each group, but additional "unrelated" information about problems experienced by women. Researchers asked men and women to draw maps of their village and the proposed changes they would like to see. The men's map reflected their orientation toward the outside world, focusing, for example, on the roads leading out of the village. Their map also reflected local politics and status, pointing out such things as the cotton trees owned by two different clans. The women's map focused on the village center and other areas of concern to their everyday lives (such as where water and fuel are collected).

Men and women then marked the maps with a series of changes they wanted to see in the village. Men lined the road leading to the village with buildings, including an administration hall, which would create a "prestigious entrance" to the village. When asked to make their own changes, the women first commented that, "Women do not have any power to decide where any of these things should be. The men have the last say." When encouraged to pretend, they drew a huge hospital and school close to the center of the village.

The maps not only clarified the changes villagers might like to see, but also revealed other important aspects of the community's make-up and culture. From the mapping exercise, the researchers discovered that two separate clans made up the village. In addition, women revealed much about their position in the village, and their concern with the health and education of their children. The mapping exercise even revealed aspects of community life that could not be seen on a map. In one village, when asked to draw changes they would like to see, the women replied, "We can't draw this map, because the kind of changes we need can't be drawn." This led to a discussion with the researcher of problems of overwork and beatings from their husbands.

Source: *Alice Welbourn. "RRA and the Analysis of Difference" in RRA Notes: Participatory Methods for Learning and Analysis, No. 14, (pp. 14-23). December 1991. International Institute for Environment and Development.*

puppet shows, role plays, songs, and drama. These lead to further reflection, discussion, and analysis.

Walking surveys

A walking survey, unlike an aerial survey, is conducted by simply walking in a project or study area. Rather than employing cameras, the walker uses only his or her eyesight and powers of observation to develop an understanding of a project area.

Walking to the periphery of a ward and then following mud tracks down into a valley may reveal very directly why the poor get bypassed in development. Walking to a water hole may show why women only fetch water once a day or switch to mud puddles when the rains begin. Walking through an area also leads to an understanding of power or caste divisions and spatial organization, architectural styles, people's use of space, environmental sanitation, overuse or underuse of facilities, and activities around water and sanitation facilities.

In the Maldives, a walking survey found that an invisible social dividing line between wards was influencing use of community toilets. Built for all people on the north end of the island, the communal toilets were in fact used only by those who lived in the ward where they were located. Questions posed to community members revealed that even young children perceived themselves as belonging to a particular ward. Although wards had been officially eliminated fifteen years earlier, the distinction was still very relevant.

In a southern state of India, a very brief walking survey was undertaken by four women from the Department of Public Health Engineering. Walking in four different directions for 1.5 hours, the women located three open wells. Official reports and discussions had not mentioned a single open well.

Mapping of communities

Maps can be made in many ways. In water supply and sanitation projects, simple maps highlighting water sources and toilet sites, primary ecological features, and settlement patterns are invaluable for planning, monitoring and evaluation. Maps can be used to understand a variety of aspects of community life: the role of women; the use of facilities by households; and social, political, or ethnic divisions. Drawn lines, thread, straw, ribbon, and rope can be used to demarcate which families use which water source or defecation sites. The spatial distribution of poor women, the rich, or leaders can be shown by using different colored dots. Many variations of mapping have been found to be effective.

In Indonesia, three-dimensional maps of villages were made by field workers using mud, clay, rocks, buttons, leaves, cotton,

Field Insight: Poverty Mapping by Stakeholders, Kenya

During a participatory poverty assessment in Kenya, the mapping activity was used at the district level, with senior district decisionmakers, NGOs, and community leaders, and at the community level with groups of men and women. In Western province in Kakamega, at a day long workshop, forty district decision makers, NGO representatives and community leaders drew district maps depicting poverty in the district. The activity included discussion of characteristics of the poor, what they did and where they lived. Work was done in small groups and color stickers were used to map concentration of different groups.

The maps of the district decisionmakers were the least detailed, while the ones drawn by the community leaders had the most detail. District decisionmakers praised the power of the mapping activity to focus their thinking on poverty issues and allocate water or health resources based on poverty criteria.

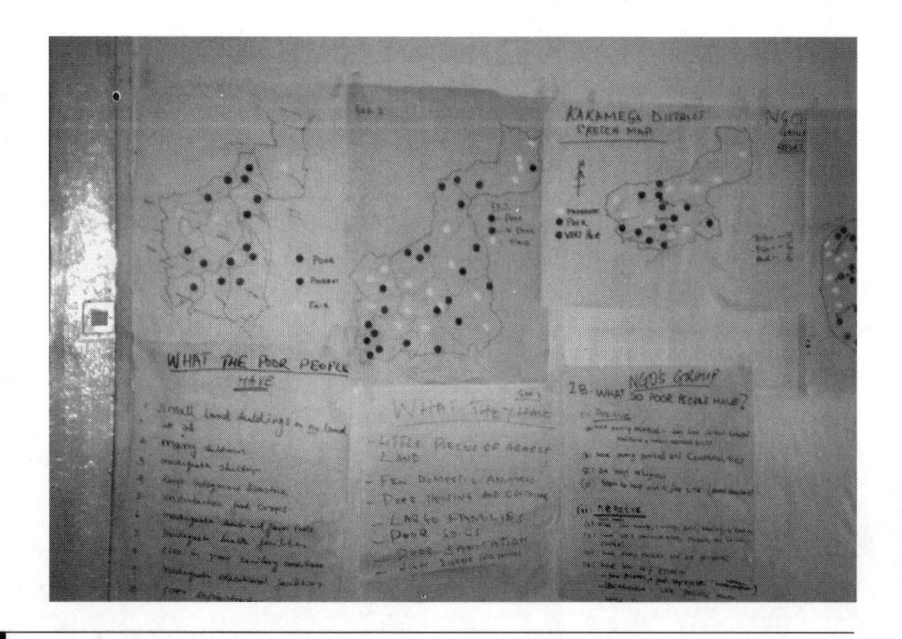

WHAT DO POOR PEOPLE DO?

(a) Spend too much time trying to meet **basic needs**.

- fetching water, firewood.
- subsistence/peasant farming
- engage in illegal activities eg. brewing alcohol
  - drug peddling
- engage in land disputes & family feuds.
- begging
- Collecting garbage, waste papers & bottles.
- hawking & petty business
- Scavenging dustbins for food (street children)

Laziness & idleness.

- too much drinking of alcohol
- smoking and taking drugs
- getting too many babies & prostitution
- fighting, gossiping.
- State of hopelessness (mainly the youth)
- Stealing, burglary & robbery.

Field Insight: Voting by Pocket Chart, Kenya

Participatory evaluation techniques can inspire enthusiasm among field workers, increase the involvement of stakeholders within the community and empower those who participate in the evaluation process. In PLAN/Embu's Final Borehole Evaluation in Kenya, men, women, children, administrators and PLAN staff were all actively involved in the data collection process. In addition, both ethnic groups in the community collaborated throughout the process. In most of the borehole areas, all users, irrespective of their position in society, contributed to the decisionmaking process, and communities successfully drew up plans of action and took corrective action.

The use of pocket charts enabled community members to conduct data collection, tabulation and analysis. Facilitators first explained the meaning of each picture on the chart. (Pictures representing domestic water supply options were used to determine which options villagers utilized and pictures of the chief, women leaders, men, women or men and women together were used to determine who made decisions about borehole operation.) Villagers were invited to take a "trial run" at voting; they were also asked their advice on how to keep the voting confidential and how to avoid voters being influenced by others. The participants decided to place the chart facing away from the audience and to have each participant go the chart to individually vote. The voting process itself was conducted quickly and other activities were developed in order to keep non-voting participants involved and interested. At the end of the process, volunteers removed the votes from each pocket in view of the other participants and tabulated them. After the process was completed, the entire group reflected on the meaning of the data and developed a plan of action to deal with the newly revealed problems. Everyone, including the blind people in the village, participated in the process.

Source: *Justin Mugo Nyagah. 1992. "Paper on the Experience of PLAN/Embu on the Application of SARAR Techniques During the Final Borehole Evauation."*

matches, paper, and other "junk." At the village level, community members were involved in drawing maps and marking roads, churches, schools, and springs. These maps were used throughout the project cycle by technicians and extension workers.

In Ethiopia, children nine to twelve years old from an urban *kebelle* in Addis Ababa mapped their communities with great enthusiasm using poster paper, crayons, and paint. Contained within their map were "danger points" for children, including bars, and many broken down toilets and uncovered pits.

In the village of Dogon Fili, in Nigeria, the entire community became involved in constructing a single community map using a two-meter length of fabric and scraps of different colors and shapes to represent the school, mosque road, houses, water points, latrines, and other local landmarks.

In Zimbabwe, community groups mapping water and sanitation facilities on paper came up with the idea of developing a cloth map, and using different colored cloth to depict different facilities. This, they felt, would ensure that the map was durable, and could be used to monitor progress.

Stories and folk tales

Stories and folk tales capture the wisdom and beliefs of a culture and pass them from generation to generation. While stories are often used in implementation of projects by field workers or health agents, their potential as research tools remains largely untapped. By revealing indigenous knowledge systems, they can convey more about the ethos of a group than statistics from surveys. Stories can also be used as a projective technique, and as a stimulus to elicit reactions, interpretations, and classification. If processed properly, stories reveal insights about a group of people that are otherwise obtained only after months of trust-building and living together.

When people feel others are non-threatening, non-judgmental, and genuinely interested in them, they often start telling stories, revealing certain myths and beliefs. Stories can also be used as a stimulus to provoke opinions or solutions, or to discuss what happened next, why things went wrong, what could have been done, who in the family made the wrong decision, and so forth. Such material can be collected by people themselves or by participatory researchers.

When stories are used to gather information and generate awareness about people's beliefs, the information collected needs to be processed either through self-evaluation or by outsider evaluation. Obviously, self-evaluation is more participatory, and leads to greater self awareness and to changing or building knowledge systems rather than mere information collection. At the end of an information processing discussion, participants are asked to analyze their responses, and reflect on what their responses reveal about their own beliefs, values, and behaviors.

Field Insight: Drawing on Indigenous Knowledge, Ethiopia

While conducting a Rapid Rural Assessment (RRA) to investigate erosion management issues and programs in a Peasant Association (PA) in Ethiopia, researchers relied heavily on the knowledge of local residents both in carrying out the RRA and in identifying problems and their solutions.

The researchers began by dividing into three sub-groups on the first day of field work to examine issues of hillside closure management, tree planting on communal land, and tree planting and management in the field and around the homesteads.

Group one was assigned the task of producing a sketch map of the area including the location of the main infrastructural components. To achieve this, the group worked closely with community participants from the area; it also used air photo mosaics.

Group two was assigned the task of assessing the hillside closure and forest land; they were assisted in this task by site guards, PA leaders and other area field workers.

Group three completed a transect through the PA, particularly noting land use, population, livestock, crop patterns, soil type, and productivity. In carrying out this study, the group talked with community participants and worked with pre-existing data from the local agricultural development center.

The following day, after the initial data had been gathered, the three research groups met with different groups of community participants to discuss findings, topical issues, and other immediate problems of the area. By listening closely to input from community members, one group discovered important communal land issues which helped shape future plans. The second group relied on information from older men and women of the community to compare current tree planting practices with previous management practices. A group discussed tree management around the homesteads with a variety of community participants including farmers not in the Producers Cooperative, Women and Youth Association leaders, and members of the cooperative.

A final meeting was held at which some participants from all the groups were present and given an opportunity to comment on the study, its objectives, and other issues. Problems and solutions were discussed and agreements were reached about a variety of management issues. Participants were frank and open about the problems, taking to heart an Ethiopian proverb offered by one of the community participants: "One who hides his illness will not get medicine to cure it."

Source: *Dessalegn Debebe. "The Role of Community Participants in RRA Methods in Ethiopia." In: RRA Notes, No. 8. January 1990. International Institute for Environment and Development: London.*

Field Insight: Gender Analysis Through Pictures, Kenya

A visual gender analysis activity in thirty-five communities in Kenya quickly uncovered what happens to a woman's assets when she is widowed, divorced or separated. The activity involved three large pictures of a man, a woman and a couple. Seventeen smaller pictures of different assets were then distributed to show various possessions owned by men, women or both. The pattern that emerged was that even though women theoretically owned their husband's land if he died, in fact the land was usually taken over by the brother or parents-in-law; when women did retain use rights, they did not have selling rights. (A number of visual general analysis techniques are described in Section II.)

In West Timor in Indonesia, people have strong beliefs about water sources — what makes a source perennial, why water disappears, who owns water — as well as beliefs about evil spirits, rainmakers, and the causes of illness. These beliefs and stories are very revealing of people's attitudes toward water and sanitation problems. Village elders will deny the existence of these beliefs to outsiders until they are convinced that the visitor is not associated with the Christian church and is truly interested in the traditional stories.

Indigenous knowledge systems

Indigenous knowledge, or "local knowledge," is unique to a given culture or society. This collective wisdom, gleaned over centuries and passed from one generation to the next, is a rich repository of cultural norms, mores, values, and beliefs. It influences every aspect of life, including people's acceptance or rejection of messages from change agents. Indigenous knowledge is a storehouse of information about successful ways in which people have dealt with their environment.

In the water and sanitation sector, local knowledge can shape behavior and practices related to the use of water, sanitation, and health facilities. At a deeper level, it can also determine how people perceive and value water, how they treat water as an infinite or finite commodity, what they believe about its origins, and whether they think that clean water can become polluted or contaminated. Tapping into indigenous knowledge systems can also reveal why local water and sanitation systems are built in ways which violate "scientific" norms.

There are various ways of uncovering indigenous knowledge systems. Among them: asking people to think of stories, beliefs, and myths about particular subject areas; talking to local experts, water diviners, traditional doctor and faith healers; talking to older men and women and children (groups that are less concerned about giving socially desirable answers); and collecting local proverbs, sayings and riddles. Visiting a school room and asking children to write down two traditional proverbs can result in a collection of more than 100 sayings within an hour. Similarly, stories told by children can be tape recorded.

Pictures and drawings

Pictures and drawings are powerful stimuli that can be used in structured, semi-structured or relatively unstructured ways. Pictures can be shown to a group of people or to individuals to elicit responses about community life. Groups and individuals can also be asked to draw pictures which can be subjected to self-analysis or analysis by researchers.

Drawings

Drawing pictures releases tensions, gets people involved, and reveals their concerns. The very fact that people draw some things and not others reveals what is of importance to them. Obviously, the content, not the quality, of the art work is primary.

In a workshop in Nepal, village women and female extension workers were asked to collect information and discuss the water and sanitation situation in specific villages. At first, the village women were dominated by the views of the extension worker. Then, the village women were left on their own, and given poster paper and felt pens.

For about twenty minutes, the women just sat, looking at each other, occasionally talking and staring at the paper. Finally, one elderly woman gingerly picked up the felt pen, stared at it, struggled to open it and tentatively drew a line on the paper. After a few more minutes she drew two houses on the top right margin, then a large pond on the bottom left side of the picture with cows walking in it and a woman drawing water. She also drew a stand post and rocks on the bottom right.

The process revealed the wide gap between problems as perceived by village women in simple drawings and the perceptions of the extension workers. The extension workers had focused on health, a catalogue of prevalent diseases, and the poor quality of water and sanitation, including lack of toilets. The village women did not once mention health, disease, toilets, dirt or sickness. They focused on the difficulty and drudgery of walking down slippery mud paths to an open pond to obtain water in which cows and animals were watered, people bathed, and children swam. The stand post was a relic from days gone by, part of a broken-down gravity-fed pipe system. The women were most interested in building a fence around the pond and somehow "partitioning" the water for different uses.

In Uganda, children's drawings of poverty were an important part of the poverty assessment conducted by the Bank.

Interpreting pictures

Pictures have long been used by psychologists in research as a more or less structured stimulus. The principle is that individuals reveal how they think or feel about a situation by whatever they perceive or do not perceive in a picture and by how they order pictures cognitively and emotionally. The more unstructured a picture, the greater the range of probable responses and the greater the skill needed in interpretation.

Pictures can thus be used to elicit information, understand knowledge systems and reveal ways in which people classify concepts, situations and their effects. Pictures can be used singly or in groups. They can be large or small; colored or black and white; line drawings or photographs; video, slides, or movies. In using pictures to gather data, the following cautions should be taken into account:

Problem of interpretation. Familiarity with the medium of pictures should not be taken for granted. Interpreting pictures is a learned skill, involving the perception of three dimensions in a two-dimensional medium. It also involves magnification or reduction of size. For people who are unfamiliar with pictures, use of pictures as research stimuli is inappropriate.

In West Timor, Indonesia, 40 kilometers from the province capital, people had difficulty interpreting pictures. For example, a drawing of a mosquito was not accepted as a mosquito because it was too large; pictures of vegetables and houses were not understood because they were too small. For some people, even after explanations, the function of pictures as representations of objects was incomprehensible.

Local appropriateness. When pictures are being used for data collection, the researcher must be sure that they are being correctly perceived. Of course, picture perception can itself be defined as part of the research task. This may be especially appropriate in evaluating health education materials developed by a project. Unintentionally comic mistakes can occur when pictures appropriate in one cultural context are used in another. Instead of assuming that objects or pictures convey the same concepts everywhere, it is important first to identify the concepts to be communicated and then identify local manifestations of the concept. In West Timor, pictures developed by an international organization in primarily-Moslem Java were pretested for nutrition and health education in primarily Christian rural areas. A picture of a Moslem traditional doctor with a skull cap and a long beard was interpreted as either "satan" or "Jesus". A picture of a dirty yard that had rotten fruit, trash (including cans) and animals wandering around was often viewed with envy as a yard of a rich man who owned animals and was able to buy canned food.

The picture of a toilet (outhouse) with a proper roof and door was perceived as a beautiful house.

Use of color. For people with little experience with pictures, too many colors can be confusing, especially when they do not correspond to reality. While colored pictures appear more attractive, they have potential disadvantages as well. For one thing, it is more expensive to duplicate color drawings. For another, colors may distract from the main message of the picture. Often, cultures have strong positive or negative associations with certain colors which, if used inappropriately, may contaminate responses.

The problem of overexposure. There is a danger of overusing the same pictures. After a while, familiarity with an image makes it less useful for research purposes. In Indonesia, a local women's organization had distributed a set of posters on cardboard paper for use by health volunteers. Because of the limited supply of images and lack of imagination, the same pictures had been used in several group meetings. When an evaluation team arrived to experiment with posters as a possible stimuli for research, it was obvious that people knew the posters too well. As a result, the pictures no longer evoked authentic responses.

Degree of ambiguity. If a research task involves finding out whether people know the "correct answer" or have some specific information, pictures can be highly structured. However, if pictures are to explore unknown areas or to discover local knowledge systems, then pictures should be ambiguous either in context, sequence or the affect they produce on the viewer.

Some pictures are more ambiguous in their interpretation than others, or involve different types of ambiguity. In Indonesia, a set of pictures was used as a research stimulus to uncover people's concepts of health. However, the pictures associated with good health tended to have "happy people" and pictures associated with "poor health" had "sad people". During pre-testing, it was found that these emotional cues were being used as a basis of classification rather than the health practices depicted. Once the cues were toned down, people's attention was not distracted by the emotional content of the pictures.

In Nepal, the force-field[1] exercise was used by a researcher to better understand the constraints and resources perceived by village women in planning and undertaking a water project. As introduced, however, the activity was too abstract for the village women. But once a few concrete examples and drawings of "constraints" and "resources" were used, the women were able to clearly identify needed resources, those that they had access to, constraints that operated at the village level and problems over which they felt they had no control. The exercise proved so successful that the workshop artist used the women's suggestions as the

---

[1] For information on Force Field Analysis, see Lyra Srinivasan, *Tools for Community Participation*, or Deepa Narayan and Lyra Srinivasan, *Participatory Development Materials Tool Kit*.

Field Insights: Use of Photography as a Participatory Research Tool

Photography can be used in a variety of ways as a tool for participatory data collection. For example, community members can observe and interpret a series of photographs presented by project staff; project participants (community members or extension agents) can take their own photographs; or the two techniques can be combined.

Burkina Faso

In a community-based pilot project for the World Bank's third urban project in Burkina Faso, photographs were used as a catalyst to increase understanding of issues among community members and extension agents. The project's extension team first produced a series of photographs which they felt showed the main characteristics of the area. This increased the capacity of the extension agents to observe the community, and facilitated their understanding of the inhabitants problems.

After the photo series was produced, the extension team constructed "photo stands" which were displayed in a variety of settings, such as in the project office, during community meetings, and during formal discussions. The photographs proved to be an excellent tool for gathering community input, and community review of the photos led to extremely open discussions. In addition, through the photos and resulting discussions, the project team and the community developed a deeper partnership for achieving the community's goals.

China

In a Ford Foundation project in Yunan Province, China, the photography approach helped to define village needs and communicate those needs to policymakers. Women in each village were given a simple camera and film to document daily life. After the photos were developed, the women were asked to develop text for each photo. Through the photographs, and with the help of the project facilitator, the women were able to articulate their priorities for village improvement. A slide show was eventually developed from the photos for outreach to higher level policymakers.

In some cases, the approach also helped to empower women to take action in resolving the problems they had identified. In one village, the photos indicated that the women's top priority was cleaning up a polluted water source. The women took numerous photos of the water source which had been polluted by a township enterprise. After this issue was identified, the women jointly wrote a letter to the mayor expressing their concern. This pressure eventually resulted in new investments in cleaning up the village water source.

Source: *condensed from Annie Manou Savina, RWSGWA and Mary Judd, RWSGEAP, World Bank.*

---

basis for new drawings of a variety of resources and constraints for future use at the village level.

Hidden biases and variety of pictures. The content of pictures predisposes certain types of responses. This bias can be decreased through the variety of pictures used and by introducing some irrelevant pictures. For example, a needs identification activity should include pictures of a wide variety of problems and contexts and not just water, sanitation, and health pictures. In Kenya, a needs identification activity using picture cards was expanded to include drawings related to different districts. Details of the pictures, such as clothing, were made specific to the regions of the country.

Degree of structure. Pictures vary in their degree of structure as well as in the way they are used and analyzed. Highly structured use includes rank ordering pictures which are then scored according to their correspondence with a predetermined correct sequence; and problems are then matched to solutions or questions to answers.

However, depending on how activities are structured, processed and analyzed, information obtained can go beyond simple "yes" and "no" answers. In a health game, participants finish sorting cards into "healthy" and "illness producing" categories, and are asked to explain their categorization. This leads to a discussion of how prevalent the illness or healthy practice pictured is in the village, and whether and why the person concerned follows the practice. All this information is recorded and subjected to content analysis using computers.

Detail and background. Field experience has shown that cluttered background or irrelevant detail in a picture detracts from a focus on the main picture or concept. Unfortunately, it often proves difficult to convince artists not to make drawings look pretty, with detailed and artistically rich backgrounds. Examining a picture of a sick child lying on a bed, mothers in a village in Thailand were captivated with the intricate pattern of the bedspread and embroidery on the pillow cover rather than with the idea of the sick child. Viewing another picture of a house with a cluttered, dirty yard, women focused their attention on the colorful window curtains. Analyzing a picture whose central concept was washing of hands before eating food, both men and women commented in detail about the food on the table in the background and whether it was healthful or not, rather than focusing on handwashing.

Photographs

Photographs are very effective as stimuli in starting discussion. Because they represent people's known environment, they capture interest and thus help recall and elicit new perspectives about a familiar situation. Depending on a program's status and philosophy, photographs can be used for planning or in a variety of other ways. "Before" and "after" photographs, for instance, can be used as evidence of change itself, and can be invaluable in aiding recall and

Field Insight: Chile

The Center for Educational Research and Development (CIDE), working in Chile, uses simulation games to involve beneficiaries in assessing the center's education programs. Simulation games — such as board games, role playing, and group discussions — are used to critically analyze a problem and search for solutions. Two examples of the more than 200 simulation games used by the center follow.

Brainstorming

Beneficiaries are divided into two groups. One group is asked to brainstorm about what went well in a project; the other brainstorms on what went wrong. Each group then creates a short skit based on its brainstorming session; the skit is then performed before the entire group. After the skits are completed, the participants then engage in discussion to analyze why positive or negative outcomes occurred in the project.

Verbal images

Used in place of survey methods, "verbal images" are obtained by asking a small group of informants to make descriptive or evaluative statements, in their own words, which present a "picture" of what occurred in a project. For example: "Participants in the education project are mostly people who cannot read or read very little." Or, "they are people with scarce resources." A series of nine or ten statements such as these is then presented to local groups of beneficiaries, who are asked to evaluate and amend the statements as they see fit. The resulting discussion generally leads to a consensus on the facts or judgments about project outcomes. Final statements from each group are presented verbatim in the final evaluation.

Source: "Who Interprets? Who Decides? Participatory Evaluation in Chile" by Horatio Walker in: *Development Communications Report*, no. 72, 1991/1.

Field Insight: Use of Visual Tools for Participatory Research

Visual tools that reflect local reality help overcome class and literacy barriers and facilitate the involvement of those usually excluded: women, the poor and the less powerful. At the agency level, visual materials help participatory modes of interaction, break hierarchical and disciplinary barriers and forces staff to explore new ways of doing things. It also demystifies planning and research. Additionally, it often marks the beginning of people realizing such materials could be used to involve community people in decision making.

Almost all materials can be used in a participatory or non-participatory way. It is easy to use innovative, visual materials to extract information from communities for external planning rather than to empower people to undertake action. Readers are encouraged to adapt ideas in the book to met their own specific needs.

In participatory activities, facilitators keep a low profile after introducing a task or activity. The tasks should be simple and the need for instructions should be minimal. This necessitates much time preparing the materials and thinking through the process. However during the actual activity, good facilitators let the process be controlled and taken over by the group to the greatest extent possible. Tasks that are open-ended allow the emergence of local perspectives, beliefs, values, reality rather than eliciting the "one correct answer".

When the intention is to empower participants, it is helpful to keep the following questions in mind in designing and conducting activities:

- Is the task open-ended or over-structured ?

- How much time and instruction is needed to clarify the task?

- Who is controlling the process?

- Who plays the dominant role in managing the task?

- Who is controlling the outcome?

- Does the task search for the one correct answer?

- Who is talking the most? (facilitator or participants)

- Does the task generate discussion, thinking, energy, excitement, fun ?

- Does the activity lead to changing perspective, group spirit or discussion of " what next" ?

A "good" participatory activity, is one in which the facilitator becomes invisible.

understanding of the "before" situation if data on the changed situation does not already exist.

In one community, enlarged photographs of a broken handpump with dry grass around it were contrasted with a recent picture of the same handpump with vegetables growing in the background. The pictures were extremely effective in drawing a group of people into discussion about how their previous situation, the problems they had during construction, critical incidents, and how water distribution was being managed.

In another project, external evaluators used enlarged photographs of an unimproved spring in a remote part of the village, and the same spring when it had been captured. The pictures proved helpful in evaluating the contact that ward leaders and village officials had with the distant ward. When the photographs were shown to the village chief, he did not recognize the spring, nor did some of the extension workers assigned to work in the area.

Photographs of extension workers, or project field workers can also prove effective in evaluating or focussing discussion on the role of field workers, how familiar they are to community members, and how local people perceive them. This is especially important in participatory projects in which the field workers are trying to facilitate and nurture the growth of local people rather than assuming leadership roles themselves. Because there may be a high turnover rate of field workers in a given area, photographs may also prove useful when interviewing community members about their contacts with specific field workers.

If a project is truly participatory, photographs should also be taken by community people. Polaroid cameras can be given to local groups of people with agreement on the number of pictures that will be taken and what the theme of the pictures will be. For example, in a group meeting with women, the purpose of the camera can be discussed and demonstrated. Women can divide into two groups according to criteria they consider significant and the purpose of taking pictures can be discussed. Participants should be encouraged to discuss the themes in their small groups and decide what types of pictures each group would like to take. If skill levels are low and women hesitate to take pictures themselves, evaluators can be of assistance, taking care not to influence the type of pictures taken.

In Lesotho, groups of women and men took turns using a Polaroid camera to photograph what they perceived to be the most important changes in the community over the past three years. Each group took three pictures. The men focused on unemployment and soil erosion; the women on new toilets and houses.

In setting up this type of process, simple guidelines should be given that offer a theme which is open-ended; do not give examples of pictures to be taken. Thematic topics might include: the most significant positive changes that have happened because of the water and sanitation project; negative changes that have resulted; problems still to be overcome; and factors, people, groups, and organizations that were responsible for the change. Different themes proposed by the groups can be further discussed, and available photographs can be made into a collage and displayed in a central place.

Photographs can also be a very useful entry point for discussion of issues that are emotionally charged and controversial, particularly those issues people avoid talking about. This might include lack of team work and problems between field staff; fights over water distribution rights; control of a facility by one or two people or households; faith in and use of traditional doctors; and female leadership.

Games and simulations

Pictures, photographs, objects, role plays, puppet shows, card games, and board games can be combined in various ways as research tools.

Prior to launching a health education program, a card activity was developed in Indonesia to better understand people's concepts of health and illness and to study how pictures are perceived by people. The activity consisted of twelve pictures of practices and situations that could be associated with good health (use of toilet, vegetable garden, garbage collection, baby weighing, eating nutritious foods, washing hands) and twelve cards consisting of pictures that could be associated with illness or poor health (flies, garbage, defecating outside, eating only starchy food). Participants were given all twenty-four cards and asked to identify the subject of each picture and then sort the pictures into two categories (health and illness). They were then asked to explain why they had sorted them as they did. The explanations led to spontaneous recategorizing and discussion of the prevalence of each practice in the village and what problems might be associated with it. The data were then subject to content analysis, with categories established, responses coded, fed into a computer and a series of cross-tabulations made to study village and gender differences.

A health education game developed by UNICEF in Java, Indonesia, was played in groups to establish people's information base about specific health-related practices such as immunization, breast feeding, oral rehydration therapy, and nutrition. The game consisted of two sets of cards colored orange and green. The orange cards had questions and the green cards had answers; each card also had a picture drawn on it. The game was played in two different ways. In the first method, a group of women sat in a circle. The facilitator shuffled the orange cards and set them in the center. The green cards were shuffled and five cards were dealt to each woman. Women then picked a question card from the center pack and tried to match it with the "correct answer". In the second method, the cards were not dealt to individuals but to two groups. Group interaction ensured lively discussion.

Data collected from such games can be analyzed by the facilitator using a simple coded sheet indicating the number of right answers. However, comments made by the group also need to be noted to understand the reasoning underlying answers. For example, in Timor, the issue of whether breast or bottle is best for an infant always led to animated discussion, although the "correct" answer was supposed to be breast feeding. Sometimes women stated that bottle milk was better because the composition was always the same, irrespective of the diet or health of the mother. Women also stated that if the mother did not have sufficient milk, or worked away from the home, bottle feeding was better.

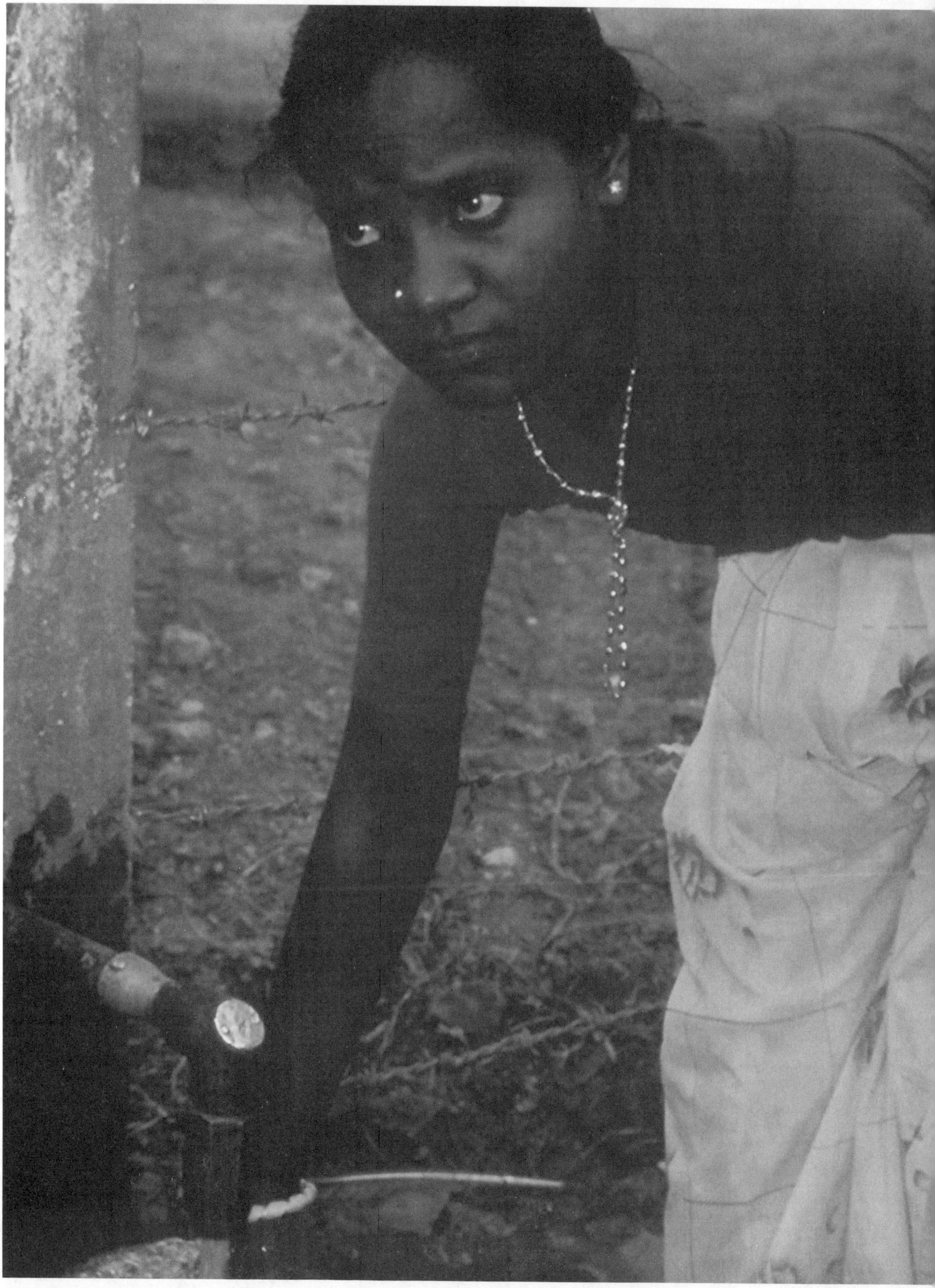

Chapter 7

Short-Cuts to Sampling

Every research study involves, to one degree or another, a sampling process. If one pump is observed rather than another, if the researcher talks to one women rather than to another, or if one agency staff meeting is observed rather than another, then sampling is taking place. Simply put, sampling is involved when any choice is made about studying some people, objects, situations, or events rather than others.

The general aim of sampling is to increase the ability to generalize results to the total population and to ensure that the sample includes units of interest to the study. Sampling techniques can be simple and intuitive, or they can be statistically defined and include formal statistically determined sampling techniques. Some people believe that formal samples have to be large. This is not true. Quite small samples can be selected using formal sampling techniques. However, when it is intended to study only a very small number of people, objects or situations, better results may sometimes be obtained by selecting them purposefully rather than randomly. Because the main purpose of participatory research is to contribute to change, action, and usable knowledge, sampling choices should be determined by agency and community stakeholders and users.

If the purpose is to draw a detailed in-depth profile of how women spend their time, a sample of three may suffice. How the three women should be selected depends on the purpose of the study. Through discussion with those who have a stake in the outcome, factors that affect time allocation in that particular cultural milieu should be discussed and consensus reached on which women to select. For example, women can be selected by age (young, middle-aged, old); by wealth (poor, average, rich); by family size (number of young children or number of other adults in the household); or by distance lived from the water source.

A question that researchers often ask is how many people should be interviewed in a study. The answer depends on the insight and abilities of the interviewer and the ability, knowledge and openness of the respondent and to a certain extent the preferences of the stakeholders who will use the data. The psychologist Jean Piaget was able to conceptualize a whole theory about the thinking of children by detailed observation of a sample of three, namely his own! One approach is to continue to conduct interviews until the type of information one gets becomes redundant or begins to fit into a pattern. There is no particular reason to talk to large numbers of people who are saying the same thing. What is important is to have some degree of confidence that the information obtained describes

Chapter Contents

This chapter explores the following issues:

- Probability sampling
  - Random sampling
  - Stratified random sampling
- Non-probability sampling
  - Accidental sampling
  - Snowballing
  - Common sense sampling
  - Quota or proportionate sampling
  - Systematic sampling
- Sampling of objects and events
- Sampling of people
- Bias introduced by drop-outs
- Determining sample size
- Choosing sampling units appropriate to the purpose
- Sampling by gender
- Sampling users and non-users

the situation as closely as possible and represents the views of a range of people.

Statisticians often refer to two kinds of error associated with sample surveys: sampling errors and non-sampling errors. Sampling errors are generally associated with probability sampling and reflect the possibility that a particular sample, even when selected using randomizing procedures, happens not to be representative of the population-at-large. Sampling errors decrase as the size of the sample is increased (so that eventually, if one were to sample the entire population, there would be no sampling error). Non-sampling errors represent all of the other sources of error that can creep into a survey. These are normally best controlled by keeping the sample size fairly small. Some common sources of non-sampling error, or biases, that should be avoided by the participatory researcher include:

- Leadership bias, i.e., sampling only village officials and leaders;

- Gender bias, i.e., sampling only men or only women;

- Age bias, i.e., sampling only the young or middle aged - excluding children and elderly;

- Visibility bias, i.e., sampling only those close to roads and village centers;

- Wealth bias, i.e., sampling only those who are well off.

Awareness of one's own biases or tendencies to choose people because they are non-threatening, easy to interview, or similar to one's self is extremely important. A conscious effort should be made to include a wide range of people without over-compensating. In a study in Pakistan, for instance, an anthropologist, conscious of his own biases, made a deliberate attempt to interview poorer people, a group usually left out of quick studies. He found that he over-compensated and had ended up interviewing a disproportionate number of the poorest people, rather than people slightly better off.

There are two broad sampling approaches — probability sampling and non-probability (or purposive) sampling.

Probability sampling

Probability sampling — selecting a sample in a way that every unit in the population has some probability of selection, and the probability is known. In order to carry out probability sampling, it is necessary to have some form of list or sampling frame available beforehand from which the sampled units can be selected. There are several ways of selecting a probability sample. Three of the most common are simple random, systematic and stratified random sampling.

Random sampling

If a study is designed to collect data on latrines, all the existing latrines are listed, the numbers put in a hat and a few selected, the method of sampling is random. A table of random numbers can also be used to select a sample. If a complete list of units under study is available, random sampling is relatively easy to do. The example of the latrines is an example of what is known as single stage sampling. For logistic and cost reasons, single stage sampling is not always practical, hence two-stage or multiple stage sampling is then used. This is also referred to as cluster sampling. For example, a number of primary sampling units, say villages, may first be selected. The second stage is selection of a sample of households within the sampled villages.

Stratified random sampling

Stratified random sampling differs from pure random sampling in that the population is sub-divided into homogenous sub-sets; a given number of units is then chosen from each group for study. The total of all the groups comprises the sample.

If latrine use is the subject under study, then the types of latrines (for example, pit latrines and water seal latrines) could form a sub-group. In this case, if the total required sample size is 150, 75 pit latrines and 75 water seal latrines will be randomly selected. The same is true for gender-related differences; if it is felt that women feel differently than men on a particular topic, the strata can be based on gender.

Stratified sampling can be more precise than simple random sampling and allows an investigator to choose a sample that represents the most important characteristics of the group. However, it may require more effort to select a stratified sample than a simple random sample.

Systematic sampling

Systematic sampling is distinctive in that it uses a more randomly structured approach to selecting the sample, but still provides a selected sample. For example, if a community consists of 200 households and a sample of fifty is needed, a systematic sample can be developed by choosing every fourth house. The starting point of the sample is determined by randomly choosing a number between 1 and 4. Thus, if 3 is chosen (by lottery), the first house to be selected is number 3; the next is number 7; and so on. If no list is available, or if existing lists are found to be incomplete, then every fourth house can be selected until the entire community is covered.

In quick surveys, systematic sampling may not be applicable, because obtaining a complete list of households, facilities or people within a population may be too time-consuming.

In any sample, there are households that do not qualify for inclusion, may be uncooperative, or are temporarily empty. To take such non-response into account, it is better to have a longer list of potential households to be contacted than is necessary. This will ensure that the process of selection is the same for all households, rather than different for some households.

Non-probability sampling

Non-probability samples include such techniques as accidental sampling, snowballing, common sense sampling, and quota or proportionate sampling. When every object, person, or event does not have a known probbility of being included in the sample, the sampling method is "non-probability," and selection is dictated by specific purpose. This is known as *purposive sampling*.

Carefully chosen, purposive samples can be extremely valuable in certain situations, especially when time is short. This is particularly true when the investigator's primary interest is in understanding qualitative and relational issues rather than quantitative problems pertaining to how much, how often, or to what degree a particular attribute or characteristic is distributed. Non-probability sampling has been found to be effective in obtaining a holistic view of a situation, and in understanding systems, behaviors, events, institutions, and underlying processes. If used in conjunction with small samples randomly chosen, non-probability sampling can provide valuable insights in explaining information obtained at a more superficial level through questionnaires or semi-structured interview schedules.

Many variations of purposive sampling of people (men, women, children), objects (water sources, defecation sites, public institutions, health clinics) or events and behaviors (water collection, use of toilets, hand washing, women speaking or being put down at meetings) can be used in participatory research. In studies requiring information on an overview of the water and sanitation situation, investigators may talk to a few "key" people. While interviewing these people can be invaluable in getting in-depth information in a short time, the tendency in many studies is to narrowly define as "key" those who are officials, project administrators, teachers, health workers and project workers. This precludes the perceptions and views of users or ordinary people. If the study seeks to understand the functioning of village organizations, institutions and leadership, it is necessary to supplement discussions with officials with lower level leaders or ordinary men and women.

There are several precautions that should be noted regarding the use of purposive sampling:

- Non-probability sampling techniques cannot be used to make definitive quantitative statements that can be generalized beyond the sample.

- Information obtained through some types of non-probability samples (such as accidental samples, below) should be thoroughly verified through other sources and methods. Data should be carefully examined for contradictions, inconsistencies, and incongruities.

- Since primary reliance is on a few individuals who are selected either because they are knowledgeable or important, or who are self-selected, it is important to understand the position of the informant in society. Thus, the age, gender, class, life experiences, occupation, family connections, and motivations of the respondent are extremely important in understanding his or her perspective and making sense of it in light of other conflicting or different perspectives.

Purposive sampling methods include:

- Accidental sampling
- Snowballing
- Common sense sampling
- Quota or proportionate sampling

Accidental sampling

When a person is sampled by accident because she or he happens to be available, or because she or he arrives at your doorstep and wants to talk, then the sampling is accidental. In a study in Indonesia, two teenagers came to the house of a study team and asked permission to speak. They proceeded to tell the team a bizarre story about why the hydraulic ram at a particular spring did not work. On checking this story with others in the community, the researchers found that, after initial denials, people acknowledged its accuracy and further elaborated upon the situation.

Snowballing

Snowballing is a sampling technique that involves asking a key informant to name other people who should be contacted by the investigator in order to understand some aspects of a situation under study.

Common sense sampling

Common sense sampling is an attempt to include a range of people or a variety of different situations in the study sample. The basic aim is to avoid error through a bias in sample by ensuring sufficient diversity. Thus, in a study of water sources, the common sense approach includes water sources located at a distance from households, poorer households or other marginal groups identified by caste, class or ethnicity, and women and children as users of water sources and latrines. One simple common sense strategy is to walk to the farthest end of a community and conduct interviews at different distances from the community center or the water source.

Quota or proportionate sampling

A researcher may be interested in the opinions of men and women, of different socioeconomic groups or of different caste groups. If the relative distribution of these groups is known, a proportionate sample can be chosen to reflect the distribution of the groups in the population. For example, if it is known that women head approximately one-third of all households, an investigator who wants to do twenty interviews might choose to interview seven female heads of households. If a community has four caste groups, equally distributed, an investigator may interview five or ten people from each caste group.

Sampling of objects and events

Information needed about numbers of water sources, types of sources, and their present condition, can be obtained by site visits. However, if ten villages must be studied and each has multiple water sources of different types, and time is limited, it may be impossible to visit every source. How can one get information that is applicable to all types of water sources without personally visiting each?

In this case, there are two basic options. The first is to obtain a list of all water sources (referred to as the universe or population) and then randomly select some sources to visit (the sample). The second option is to first establish a list of all known existing sources and then, depending on purpose and time limitation, visit each.

Option 1

In a study of handpumps, for example, if a complete list of all 500 handpumps in the communities to be studied is obtained, and 50 handpumps are randomly selected, then the findings can be generalized to the population of 500 handpumps. In a random sample, especially if the sample size is below 30, the sample may be too small to conduct any statistical analysis.

If it is felt that the presence of caretakers has a strong influence on breakdown rates of handpumps, a stratum or layer is added to the sampling process. One can make two lists: handpumps with caretakers and handpumps without caretakers. Within each stratum, a sample can be randomly selected (stratified random sampling). If twenty communities comprise the sample for a study of water sources and it is known that there are multiple types of sources in each village, it might make sense to choose four different types of villages and sample all sources in each village. It should be noted that each of these methods presumes that information about all existing sources is already available or will be readily available in the communities.

Option 2

The second option is to establish a list of some of the existing sources and then, depending on purpose and time limitation, conduct a site visit to each source. An alternative is to talk to the people who are knowledgeable about the sources. In this case, it is still important to visit a few sites to understand and interpret the situation described by others.

Depending on the purpose of the study, it may be desirable to stratify the sample. To study the influence of caretakers on handpump maintenance would require making two lists — handpumps with caretakers and handpumps without caretakers. Within each list, the handpumps to be studied can be randomly selected.

However, if it is not feasible to identify which handpumps have caretakers and which do not, the separation of handpumps can still be done at the stage of analysis to see if any differences exist between the two groups. Unfortunately, in many situations, complete or accurate lists are not known immediately even in the communities.

In a study in Indonesia, existing lists of water sources located in villages and information obtained from a pre-survey were used to design the sampling framework for a survey on water sources. The pre-survey information was compiled by talking to village officials. However, after three days of research within the communities, researchers found that wards that were reported to have only two sources had as many as sixteen sources.

In a study in Indore, India, government staff and village officials reported that there were no step wells in the community. A quick two-hour survey revealed the presence of two step wells in individual compounds, along with another well in the area of common lands that are used regularly.

Sampling of people

A study of even a small community of fifty households with a population of 300 is probably too large to include everyone in the sample. Once again, depending on the purpose of the study and whether statistical tests will be used, a sample of people will have to be chosen. In selecting a sample, it is important to ensure that a wide range of people have been included to represent the diversity of opinions, attitudes, practices, and socioeconomic and demographic characteristics.

Bias introduced by drop-outs

If only those people who happen to be home during a research visit are included in the sample, it can bias or distort the study's outcome.

Poorer women may be in the fields doing agricultural work at the time of the visit; in some societies, many households move temporarily to farming areas during the peak agricultural seasons. If these people are not included in the sample, the sample is not representative of the total population. If, on the other hand, a family is away temporarily at another village, or if people are sick, an investigator may not be able to wait until these people return home or get well. Thus, clear rules should be established beforehand to determine when it is important to revisit a household for data collection and when it is not. In either case, it is important to keep track of non-responses.

Determining sample size

The required sample size can be decided with statistical precision depending on how concerned one is with sampling error, and the degree of confidence desired about the representativeness of the sample. Two factors influence the choice of sample size: (1) The level of statistical precision desired (the larger the sample the smaller the sampling error), and (2) the variability of the universe being measured.

Rules of thumb in determining sample size

- Whatever the sampling technique used, it should include a range of the units under study (types of toilets, water sources, officials, non-officials, men, women).

- If required sample sizes are large and time is limited, it is useful to study in depth at least a small proportion of the units in the sample. A brief interview schedule could be administered to 300 people, while in-depth interviews or observations could be conducted in thirty households.

- Random sampling may create suspicion among people who do not understand either why they were selected or why they were left out. One way of overcoming these problems is to publicly explain how and why the sampling is done. One can go a step further and select names by holding a lottery in public. A preferable selection method is to have a child publicly pick names from a can, since no one can accuse a child of being biased. Such a public event not only creates understanding and interest in the study, but can help develop trust within the community for the researcher and the entire data collection process.

Choosing sampling units appropriate to the purpose

A study can have a single sampling unit or different sampling units for research within the framework of the larger study. Careful attention must be given to the choice of sampling unit. Commonly used sampling units are villages, hamlets, and households.

Within villages, many studies use households as the unit of sampling and analysis. Because the definition of a household varies widely from culture to culture, it is important to operationalize this definition in any given study. This means that the researcher must use certain criteria to delineate the boundaries of a household. In many countries in West Africa, several families live within one compound. Determining whether each hut or the entire compound will be designated as a household is key to the study.

In studies which use the household as the sampling unit, decisions will have to be made on who provides information. Is it the household elder, an adult male or female, or household head? These decisions should be given careful consideration, keeping in mind that there may be strong gender differences in knowledge about and interest in water, health and sanitation issues.

Sampling by gender

When people are chosen based on gender, additional decisions must be made about characteristics that define the unit. Will any male or female qualify for the unit, or will other characteristics be used to define the type of male or female to be interviewed. In one study, a female had to be between seventeen and fifty-five years of age and had to be a long-time resident in the community (over five years) to qualify for a sample.

Sampling users and non-users

Many studies are interested in finding the differences and similarities between those who use certain water sources or latrines. The problem arises in attempting to define who are and are not users.

Imagine a study focusing on rehabilitation of handpumps, in which the investigators wants to conduct stratified random sampling of handpump users and non-users. In a preliminary survey, they ask people directly "Do you use the handpump?" Everyone responds affirmatively. The problem is that the researchers have no idea whether people actually do use the pump, or whether the respondents are simply saying that they do something that they know they are supposed to do.

Similarly, assessing latrine use by asking people if they use the facilities is likely to result in the inclusion of some non-users in the positive responses. In a study in Indonesia, more than 90 percent of people interviewed said that they used latrines. Observations during village stays proved that the majority of those saying yes were telling interviewers what they perceived they wanted to hear. (See Chapter 5 for suggestions about how to approach sensitive information in interviews.)

Another problem in assessing use, whether of vendors, public standposts, handpumps or latrines, is that a comparison group is

Field Insight: Interviews Yield Sensitive Information, Cameroon

Interviewing a variety of stakeholders is an important aspect of gathering data, especially when "sensitive" information is concerned. A team evaluating the CARE Potable Water and Community Health Education Project in Cameroon designed separate questionnaires for distinct groups. These stakeholders included community leaders (such as the chief, president and members of the Village Development Committee); the men and women who were responsible for pump maintenance; the Sanitary Action Group (GAS); and village women.

By using a female team member to interview the local women, the team soon had access to a wealth of new information, even in areas which they, as outsiders, had not previously considered to be "gender sensitive." For example, the women's responses during interviews conducted by a female evaluator (without the presence of men) did not correspond with the responses of the Village Development Committee or the Sanitary Action Group. The village women maintained, for instance, that health education did not exist, while the GAS in several villages stated that education was part of their activities. Women often told the female evaluators that they did not know the role of either the GAS or the Village Development Committee, and in four of the nine villages they were unaware of the women on the project committees.

Source: *Mark Baron, Marcel Zollinger, and Clarissa Brocklehurst. 1989. Rapport D'évaluation: CARE Cameroun Project D'Eau Potable et D'éducation sur la Santà Communautaire Province de L'Est. Préparé pour CARE Canada par Cowater International Inc., Ottawa, Canada.*

needed. This may seem obvious but is often not done. A study that samples only handpump or communal latrine users cannot make definitive conclusions about why people do not use a handpump or latrine. The study must include non-users as a comparison point.

A few simple techniques can be used to overcome these problems:

- Choose an informal random sample that includes some users and non-users. The categorization into two groups is done at the stage of data analysis, rather than at the beginning of the study.

- Because distance affects use, especially of communal facilities, choose some people who live close to public water systems, standposts, kiosks, latrines and pumps, and some who dwell far away.

- Use a snowballing technique and combine it with the sample obtained through observation. If one or two households or people are definitely known to not use a facility, ask them if they know others, or observe alternative water sources or defecation sites. Through these observations, small samples can be obtained which will correctly identify the nonusers.

Studies of non-users are often more revealing in identifying a way forward. In Bamako, Mali, focus groups with people who were throwing household garbage just outside their compounds was very revealing. People said that while they valued cleanliness, they had given up walking to the public garbage bins which were overflowing and from where garbage was not collected for days. For them, the extra time it took to walk to the public garbage bins, only resulted in garbage accumulating in another close by place and hence was viewed as not worth the trouble.

The fit between purpose, data collection techniques and sampling

It is important once again not to forget that choice of sampling techniques and sample size depends on the purpose, data collection techniques, number of field workers and resource and time constraints.

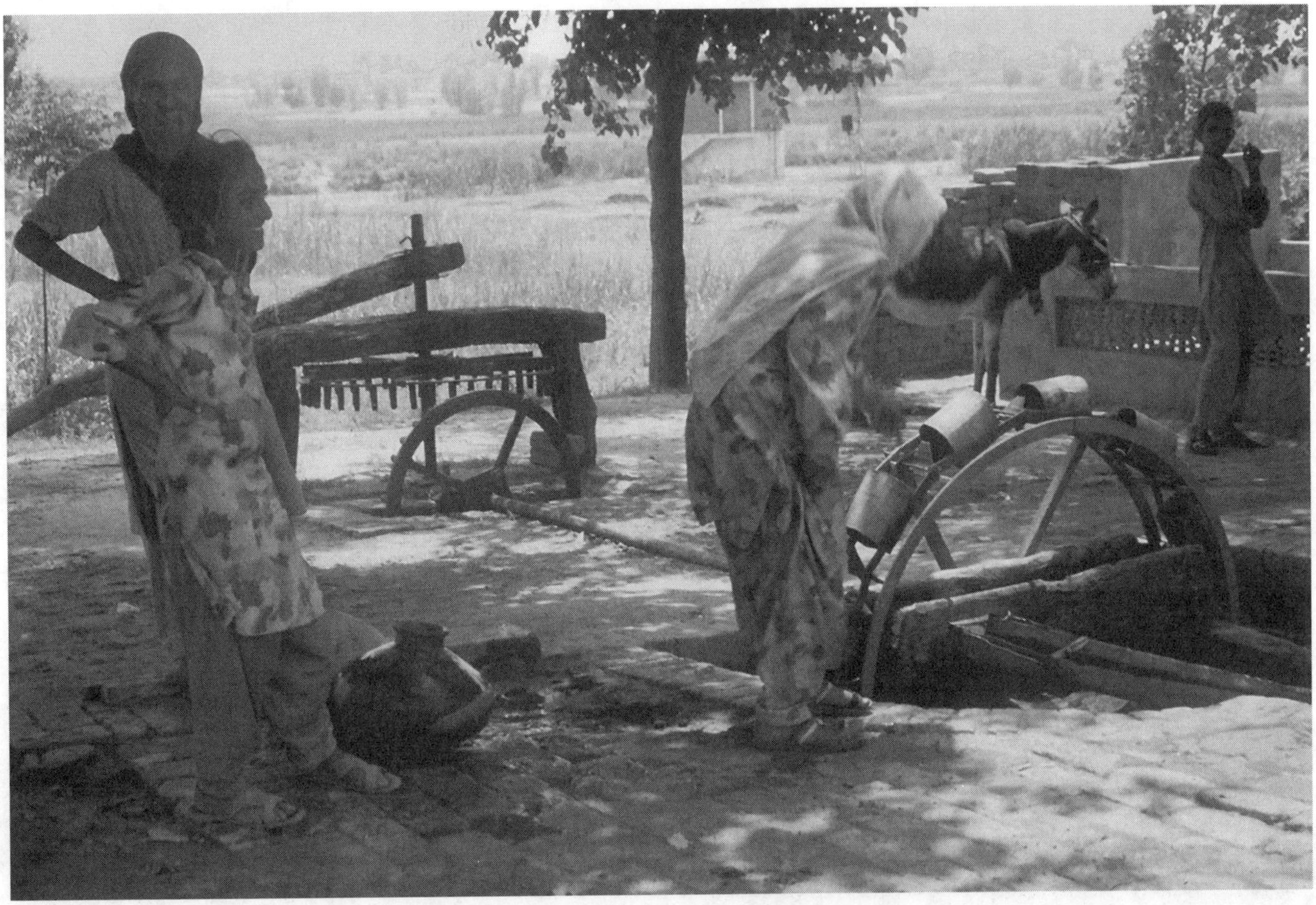

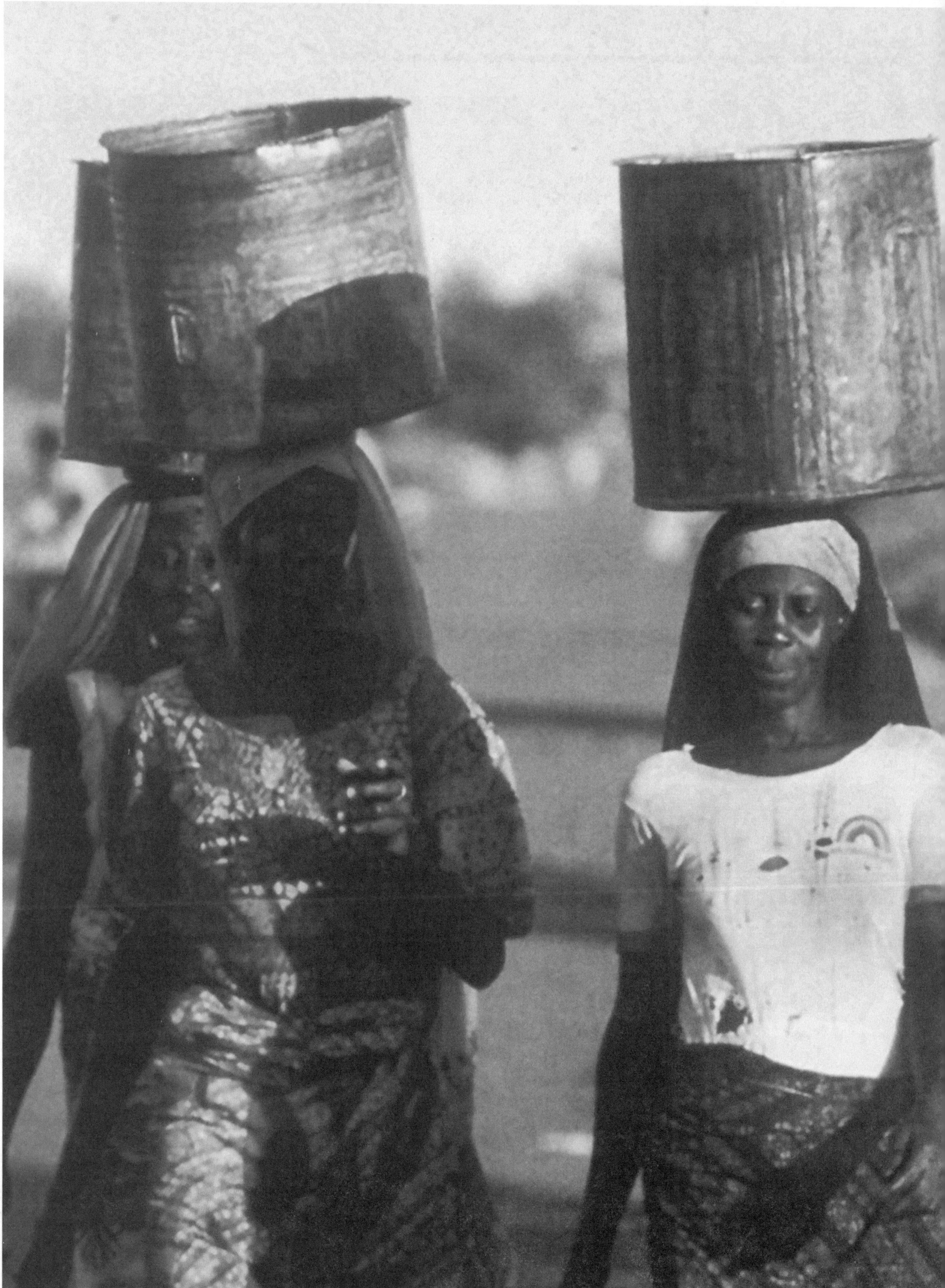

Chapter 8

Selection and Training of Field Workers

In large projects, as well as in small projects in which much data must be collected quickly, research needs to be undertaken by more than one person. Whoever collects the data — community people, extension workers, technical staff or university students — will need to be trained in participatory and short-cut approaches. The nature of information, the degree of desired accuracy, its sensitivity, the data collection techniques to be employed, and the time and resource constraints all influence the decision about who should do the collecting.

For example, if only men are available for data collection and they have no research or rural living experience, it may limit who they can talk to (not to women if the information is "sensitive"); what they can talk about (not about kitchen hygiene); or how they can gather information (not observing water points where women bathe).

Clearly, if observational data about kitchen hygiene or water use at source are needed, women must be involved in data collection. Alternatively, if such data are not needed in great detail or accuracy, then interviews with women may be more appropriate than posting male observers in the kitchen or at the water source. Other options are also available, including use of local people or school children as observers.

The fit between data collection techniques and who collects the data is important. If the human resources available for research are fixed and cannot be changed, then techniques should be chosen to ensure that they can be implemented by the number of people who are available. In all other situations, the issue of who collects data should be based on the type of information, accuracy desired and methodology chosen. Constraints have to be kept in mind. Often it is necessary to make compromises, given available resources and, if necessary, prioritize the importance of different types of information.

The success of the data collection process depends on how field people are trained, and particularly how motivated and self-aware they are. Other factors are of secondary importance, such as education, previous training, and skills in using various data collection tools such as checklists and observation schedules. Regardless of education or training level, experience indicates that the ability of motivated people to serve the function of data collectors should never be underestimated.

For ease of reference in this chapter, people involved in data collection will be designated as "field workers."

> **Chapter Contents**
>
> *This chapter explores the following issues:*
>
> - Who is a good field worker?
>   - Insiders vs. outsiders
>   - Gender
>   - Education
>   - Language
>   - Culture
> - Who collects the data?
> - Forming teams
>   - Specialization within the team
>   - Professionals
> - Training field workers
>   - Raising motivation through participatory training
>   - Self-awareness

Who is a good field worker?

A good field worker is one whom people are likely to trust, who is perceived as friendly, is a good listener, is non-judgmental and most important, does not exhibit feelings of superiority. Field workers should know the language of the people with whom they work. In selecting field workers, their individual characteristics as well as their effect on the composition of the team should be carefully considered.

Some important factors that should be examined as part of the selection process are:

- Insiders vs. outsiders
- Gender
- Education
- Language
- Culture

All of these factors need to be considered in relation to the overall purposes of the study and the data collection techniques to be used. Even when all data collection is done by community people, most of these factors are still applicable because no community is homogeneous.

Insiders vs. outsiders

The issue of "insiders" and "outsiders" is central to conducting research, whether through conventional or participatory approaches. Both can make use of insiders and outsiders to reach their stated objectives.

Recruiting Participatory Researchers and Trainers

In selecting participatory researchers and trainers, candidates can be asked the following questions:

1. Where would you wish the learning to take place? (If the candidate responds, "mainly in classroom," reject the candidate. If the candidate is opposed to sleeping in villages, reject. If the response is "mainly or entirely in village, very rural or poor urban areas"... OK so far.)

2. What do you consider most important in the learning experience? (If the teaching and correct learning of methods, reject. If behavior, attitudes, do-it-yourself, being taught by villagers, learning to unlearn, learning not to interrupt, spending overnight in villages... OK so far.)

3. What do you feel about others sharing in the training, as resource persons and co-trainers from other organizations, and as participants from other organizations? (If they want to keep it to themselves, reject, unless they are known to be highly experienced already and have a good reputation. If others are welcomed... OK so far.)

4. What hours would you expect to work? (If 0900 to 1700, reject. If starting early, and all hours and into the night... OK so far.)

5. What is your view of the participation of women and poorer people? (If considered secondary, reject. If given priority... OK.)

Try to assess the behavior and attitudes of would-be trainers and researchers. A good trainer is likely to be relaxed, sensitive, democratic, a good listener, with a sense of humor and fun, and blessed with physical stamina.

Do not recruit (a) middle-aged men who wear suits and polished shoes in the field; (b) fashionable radicals who spout about sustainability but are scared of sleeping in villages; or (c) people who take themselves more seriously than their work.

Source: *Robert Chambers. 1993. "A Note for World Bank Staff on Participatory Rural Appraisal."*

Use of insiders. Insiders are people who belong to the organization, department or community that is going to conduct the study or that is being studied. This broad definition includes senior personnel from ministerial levels, project staff, and men, women, and children from target communities.

If it is anticipated that outsiders (that is, strangers) will have problems being trusted and establishing rapport with the agency, group, or community being studied, then the obvious solution is to use insiders, people from within the group or community, for data collection. Insiders have advantages in terms of being easily accepted and able to correctly interpret and attach cultural meaning to events and situations. In a study in the Philippines, for example, field workers recruited from barrios proved more effective than outsiders in gathering information through household interviews.

Perhaps the biggest advantage of using insiders for data collection has less to do with the quality of data generated than with increasing the probability that the results from the study will be accepted and utilized by all categories of users. In a study in an island nation in Asia, inclusion of senior planning staff from the Ministry of Health as data collectors changed skepticism about "research findings." It also resulted in the study results being extensively used to design a national investment program in water and sanitation.

Field Insight: Community Women as Researchers, Ecuador

An environmental and occupational health risk assessment carried out by the WASH and PRITECH projects in peri-urban areas surrounding Quito, Ecuador shows how participatory ethnographic field techniques can break assumptions, yield new information, and create a better project. These methods included focus groups, individual in-depth interviews, and structured observation of homes and communities. Sixty women from three peri-urban centers participated in focus groups which were coordinated by two local women. These coordinators recorded, analyzed, and discussed the findings with the project team, often helping to alert the facilitator to follow a lead provided by the participants. The involvement of the local coordinators proved crucial to the success of the project.

The information women provided through this participatory process proved invaluable in changing the team's initial assumptions and broadening their understanding of the issues. For example, the project team noted originally that the entire community had access to a system for sewage disposal which appeared adequate. If they had not involved the women, the project team might have ended their dry season visit thinking the system was operating adequately. However, one woman told the project team, "All is fine right now, but wait until the winter when it rains and the lagoons return." This comment then initiated discussion among all the women, who began talking about the problems in the rainy season when sewage backs up and overflows into houses and streets.

In addition, the team's focus on water storage and occupational health issues was dramatically changed by the women's participation. Prior to the research, the team had placed importance on the problems of hygienic water storage, but not on water access and transport. Through interviews and observation, the researchers discovered that the publicly provided water (delivered by trucks) did not always reach the peri-urban areas, and many families had to rely on private truckers. Since these truckers were not subject to municipal regulations they were more likely to deliver contaminated water.

Source: *Linda M. Whiteford. "Women's Voices Heard in Ecuador Health Risk Assessment." in Voices from the City: Newsletter of the Peri-Urban Network on Water supply and Environmental Health Sanitation, vol. 3, September 1993.*

> **Field Insight: Teenagers as Observers, Angola**
>
> One of the many questions which must be answered in data gathering of all kinds is: Who should do the gathering? An innovative approach was utilized for a study of women's time saving and changes in water use in a province in Angola. The study took place at a UNICEF-assisted water program which had been in implementation for some time.
>
> Study designers proposed a time-budget survey using observation techniques in which women's activities would be timed throughout the day. The designers recognized, however, significant problems existed with this type of approach: few investigators can spend a whole day with rural women, and the presence of an investigator is also likely to affect the family's behavior. To overcome these obstacles, the designers decided to have the teenage children of the household (ages 14 - 18) conduct the investigation.
>
> Involving the teenagers of the household, rather than external investigators, had obvious benefits. These included the likelihood that there would be less disruption to the families' normal behavior patterns; that there would be fewer cross-cultural misunderstandings; that the children would be available however early the women began their day or however late they ended it; and that the sample size could be much greater. The involvement of teenagers would also be a useful educational experience for them. Because of the decision to use teens to collect the data, the sample was only representative of those women with adolescent children.
>
> Source: *Valerie Curtis. 1988. "Evaluation of UNICEF Assisted Water Programmes in the People's Republic of Angola." London School of Hygiene and Tropical Medicine.*

However, in the same example, problems also resulted from the participation of senior government officials. When senior officials have strong personal agendas or political aspirations, their inclusion in data collection teams becomes a double-edged sword. In the example mentioned above, one of the senior planners with political connections insisted on holding relatively large group meetings to ensure he received maximum exposure to people in the villages. The large meetings were always less effective in terms of data collection than were the smaller group meetings. If senior officials refuse to work once the study is underway, researchers find themselves helpless.

There are other potential problems in using insiders, as well. Insiders may find it difficult to break away from their traditional roles to establish their new identities as field workers. They may feel more pressed into participant roles, leaving little time to function as field workers. They are also more likely to be drawn into and be affected by local politics and rivalries. They may also miss significant events and patterns which they perceive as ordinary, normal or inconsequential. Some studies which have recruited local volunteers for data collection have experienced problems. A study in India found that village women did not like to give information they considered private or thought may potentially be used against them to someone from their own community.

These problems notwithstanding, there are many beneficial ways to use insiders in the data collection process. Insiders can play an invaluable role in helping to interpret or explain data. Local people can be used for certain types of data collection, especially in observation studies. In Bangladesh, Indonesia, and Kenya, local illiterate women have worked effectively as observers for water collection activities. If the community is informed and approves of the idea beforehand, children can make keen observers as well as excellent interviewers.

In a study about prevalence of diarrhea and its treatment in rural Zimbabwe, school girls who interviewed their mothers and neighboring women reported that the use of traditional medicine was 50 percent higher than when outsiders interviewed mothers. Clearly, this was information that the women did not feel comfortable sharing with strangers.

Insiders as participants. The value of insiders as data collectors when a study is viewed as a step in stimulating community involvement is indisputable. When people are involved in deciding that it is necessary to collect information, what information should be collected, and how it should be collected, then data collectors will be trusted and information will not be withheld or falsified. In addition, the process of gathering information increases people's understanding of existing situations.

In a training of community workers in Nepal, workshop designers gathered some information about villages from secondary sources and developed a curriculum for the workshop. After the first three days of the workshop, however, the workshop leaders handed over the responsibility of designing the training to the community workers. The community workers immediately became more animated. Gradually they were able to move from information needs to identifying their own strengths and weaknesses, finally developing a curriculum for the workshop which the external resource people then followed.

Use of project staff. Project staff represent an important category of insiders; the advantages and disadvantages of using project staff for field work should be carefully considered. The advantages are many. Involving project staff in data collection sensitizes them to issues and increases the likelihood of results being utilized; it also results in building staff skills in data collection that can be drawn upon again in the future. There are also obvious cost savings in using staff for field work. In addition, involving staff people may result in greater control over their availability.

There are potential disadvantages as well that are particularly related to the timing of data collection in the life cycle of a project. If project staff are used for data collection prior to implementation, disadvantages are minimal. However, if project staff have been closely linked to project implementation, two problems can arise. The first has to do with the difficulty that project staff, who may identify closely with the program, have in hearing criticisms of the project. They may also find it difficult to separate their role as "educators" from their role as "learners." In a nutrition and hygiene education project which was being linked with water supply and sanitation, female project staff were initially used for data collection. However, despite training, many of the staff found it impossible to remain in the role of interviewer. When women showed little evidence of nutrition awareness during the interview, the staff proceeded to "educate" them.

The second problem is how the community perceives the field workers. If the person involved in data collection is also the person who was identified with the water supply or sanitation and health situation at the community level, people may not feel free to be critical for fear of giving offense. The probability of getting answers that are designed to please the interviewer is high. This dynamic, however, can be overcome by establishing a reputation for openness.

Gender

Most studies need to collect information from women because women are usually centrally involved in household water and sanitation management. For people who come for short periods from the outside and hope to gain insights into the lives of men and women, it is essential that women field workers talk to women and men talk to men. Men are not likely to be very pleased if male field workers visit their homes to talk in private with their wives while they are away. Women will often be less trustful and open with strange men than they are with a female outsider.

In one study in India, a team leader insisted that an all-male team had successfully been able to talk to women because, he said, "In these villages women are not shy, they are quite advanced." Of course, men can talk to women! But the point is that men will not get the kind of information that may be given to a women interviewer. A woman is very unlikely to tell a man about problems she has in using the bush or latrine, especially during diarrhoea, pregnancy, or during menstruation. Therefore, a field team must include women.

Women are often excluded from data collection teams because it is easier and more convenient to get men to go out to a village or community. However, the cost of such convenience is high in terms of quality of information gathered.

Education

The importance of formal degrees or previous training of field workers is overrated; it is not necessarily true that people with university degrees make better field workers. In fact, highly educated people may make poor data collectors because they jump to conclusions too quickly. Often, they assume they understand a situation and that they are more knowledgeable than ordinary people living in slums or villages. Coming from this perspective, it is easy to misinterpret events or a situation; it can also be difficult to disentangle perspectives of the person interviewed or situation observed from the perspective of the field worker who is adding his or her interpretation of the situation.

One study on sanitation included some field workers who were university-educated and others who had only a grade six education. Some of the university-educated field workers took longer to learn not to summarize and to exclude their interpretation from field notes than those with less education. For example, one of the less educated field workers recalled nearly verbatim a woman's responses to his questions from his notes of the interview: "I don't like to use the community latrine... . The doors have become sticky .... Once I was stuck inside, it was dark and I could not open the door.... I feel shy because everyone can watch you.... The walls are low and you have to come outside to get water." During the same interview, one of his more educated colleagues did not take notes, relying instead on a broad — and incorrect — interpretation of these remarks: "Women are afraid to use community latrines because they are afraid of the dark and that they will be seen."

In a study in Kenya which included several teams headed by senior research supervisors, the most experienced anthropologist submitted the weakest reports, whereas other social scientists considered less experienced worked harder and produced reports with greater ethnographic detail.

Language

Field workers should be fluent in the language of the people they seek to understand. However, when this is not possible, interpreters can be used with care. Interpreters can be either outsiders or insiders, depending on the status and confidence of the people being studied. For example, use of external interpreters may not be viewed as threatening by a village head, an extension worker or a doctor at a clinic. However, the same outside interpreter may be threatening to ordinary community residents. In such cases, it may be important to choose "neutral" local people as interpreters to avoid overwhelming members with strangers.

Interpreters need to be trained in their role of serving as a "mouthpiece" without adding to questions and answers, subtracting from them, or clarifying them on their own. To avoid a situation where the conversation flows between the interpreters and respondent while the field worker loses control, it is useful for the researcher to take the "focal" seat facing the respondent and the interpreter to be positioned to one side. This will indicate each person's relative importance in the process.

Field Insight: Insiders Know the Culture, Morocco

In the World Bank-funded Socio-Economic Study of the Rural Water Supply Project in the Ziz Valley and the Tafilalet Plain in Southeastern Morocco, local young women trained by project staff collected a variety of data from fifty-four villages. The study used a series of questionnaires to collect quantitative data from the total sample of 647 households; qualitative data were collected through a series of overnight observations of fifty-four families.

The eleven female researchers, who worked in agriculture and health, were already experienced in dealing with sometimes-secluded local women. Recruited by the local water officer and interviewed by the project director, the researchers received both classroom and on-the-job training from project staff. (On-site training was completed in villages which were not in the final sample group.)

The young women proved to be an excellent choice for conducting the data gathering. All of the researchers spoke Moroccan Arabic and some also spoke Berber. They knew the area and were comfortable with the idea of spending the night in rural conditions. In addition, they were culturally sensitive to issues such as local customs of hospitality which might arise during the overnight stays. The young women also used their knowledge of local practices to help ensure that the sample of households was representative. For example, on the issue of wealth ranking, one local researcher pointed out the importance of noting whether the family had cloth mats or foam on the floor for sitting and sleeping.

Source: *Susan Schaefer Davis. 1993. "Etude socio-economique du projet d'AEP des Zone Rurales de la Vallee du Ziz et de la Plaine du Tafilalet." Vol. 1; and personal communication, Feb. 1994.*

Culture

Culture is a complex of values, norms, mores, and beliefs that influence the behavior of a group of people. Each of us is brought up in a particular cultural context; we internalize to a greater or lesser extent the values, beliefs and norms of the group in which we were raised.

When two people are engaged in a conversation, the interaction is influenced by the values and beliefs of each person. How something is said or interpreted depends on the cultural meaning that is attached to it. Hence, when a researcher is trying to understand what she or he hears or observes, it is extremely important that the person has a high degree of self-awareness. This allows automatic interpretations to be set aside based on one's own cultural experience in favor of interpreting information from the cultural context being studied.

For example, in a village in central India, an older man tried to explain during an interview why he was building a new latrine and not using the latrine recently built by the project. The old latrine, he said, did not face the "proper direction." The question of "direction" was not pursued by project staff but an outsider expressed interest in the issue. In response to additional probes, the old man said "The toilet does not face the east. It is very impolite to put one's back to the sun. Do you greet guests with your back towards them? [The old latrine faced the west.] How can you do surya namaskar [greet the sun] early in the morning with your back to the sun?"

Who collects the data?

The question of who collects the data is always of central importance. In a participatory approach to research, the implications, advantages, and disadvantages of different types of people from within the group gathering different types of information need to be discussed. Additionally, local divisions based on tribe, religion, caste, class, and occupation, and their implications for data collection, should be examined. Once the qualities of good data collectors have been discussed, the community's decisions about who should collect data must be respected and not altered under any circumstances except at the initiative of the community.

In a village in Timor in Indonesia, community members discussed the impact that staying in particular types of households would have on the degree to which outside researchers would be accepted by the community. Villagers made it clear they would not talk freely to the researchers if they only stayed with the village chiefs. Conversely, if the researchers stayed only with the poorest families, they would be viewed with suspicion and lose credibility.

In another area of Indonesia, communities chose village members to observe water collection. The external research team made an

---

Field Insight: Establishing the Cooperative Spirit, Kenya

In its Maasai People's Program in Kenya, World Vision International used community motivators (CMs) fluent in Maasai language and culture to help each community to participate in the participatory evaluation process.

In the first stage of the project, the CMs used a variety of techniques to promote cooperation and discussion. In one game, "secret in a box," community members divided themselves into three groups to discover the contents of a small box. The first group was allowed only to shake the box, the second to feel the contents, and the third could look, but not touch. In one of the first communities involved, the game helped people to relax and express opinions, and spawned serious discussions about the importance of involving the whole community in identifying local problems. Another game — "the broken squares" — in which community members divided into groups and worked together to form squares of equal size — helped establish the importance of each member's participation. In the village of Olengomei, the children finished first, the women second, and the men last, emphasizing that even "weaker" members of the community have something important to contribute.

Source: *Daniel Ole Shani and Mitali Perkins. 1991. "Participation in Development: Learning from the Maasai People's Program in Kenya." World Vision Staff Working Paper No. 12.*

Field Insights: Tips for Research from Nepal, Indonesia, Sri Lanka and Thailand

Following are some lessons presented by participants at the WHO-PROWWESS Third Inter-Country Workshop on Case Studies of Women's Participation in Community Water Supply and Sanitation. These insights were gained during three years of field research in Nepal, Indonesia, Sri Lanka and Thailand.

- Size of the project, type of data required, and time and budget constraints determine which method to use.

- Quick surveys are useful for gathering information needed immediately by the implementing agency. Follow up with more qualitative, in-depth probing to check reliability.

- To effectively cross-check data, the same issues can be explored using different techniques, researchers and respondents.

- Close-ended and direct questions are extremely ineffective when trying to obtain socially sensitive information.

- In countries where dissent is culturally inappropriate, focus groups involving leaders are largely ineffective; once male and female leaders speak, other participants tend to agree since it is impolite to create conflict.

- Researchers should be sure to be appropriate to local standards and norms when preparing to collect data in a community. Approach communities through established authority structures; use letters of introduction; get necessary clearance; inform headman of visit; clarify the purpose of the visit; avoid raising false expectations; go through necessary rituals/ceremonies; accept local hospitality; practice reciprocity.

- Potential obstacles to data collection include: villagers do not want more studies; language barriers; households scattered over large areas; severe time limitations imposed by planners; rainfall affecting access to a community; villagers unavailable due to seasonal migration, seasonal work schedules, or local time cycles.

- Characteristics of a good field worker: motivated; fresh, inquiring mind; friendly, intelligent and socially sensitive; should understand the purpose of the research and the proposed intervention; literate or illiterate (illiterate people can be used in participatory data collection and observation).

- Pre-testing should be used to screen out ineffective field workers.

- Instant checking and feedback is often necessary for field workers. Field workers should make corrections and collect missing data the next day.

- Monetary incentives or punishments are usually divisive among field workers.

- Field workers need supportive supervision which encourages improvement and keeps morale high.

Source: *Final Review of Case Studies of Women's Participation in Community Water Supply and Sanitation. Report on Workshop, WHO and PROWWESS. Kupang Indonesia, 23-27 May, 1988.*

argument for selecting women to be water point observers since water collection was primarily a woman's activity. Each observer would be given a small stipend. After much heated debate, eight young men and only two women were chosen. By the end of the second day of observation, all but one of the men had dropped out. The community met again and finally selected women and young girls to be observers. Although data and data collection time were lost, the fact that the community decision was respected meant that when problems wusually an advantage at the field level. If participatory approaches are applied to data collection, then diversity in background is a distinct advantage.

In a quick study in Indonesia (a series of village visits lasting between two and seven days), a research team consisted of men and women who were students and university faculty from different departments and different parts of Indonesia. Their ages ranged from eight to more than fifty-five years. Each person was used for tasks in which they excelled, such as leading a group meeting, training of local women for observational data collection, household interviews, or participatory mapping of the community or playing a health game. The agriculturalist spotted the use of DDT for home sanitation; a younger women was able to accompany women up and down hills for water collection; an older team member's presence established credibility with village chiefs. Some of the team members were originally from areas similar to the villages being studied. They could not ask questions about many aspects of life in the area because it was presumed that they knew the answers. However, outsiders who came from other parts of Indonesia could freely ask questions that an insider was presumed to know. A foreigner could ask even the most simple questions without appearing to ridicule the people because it was presumed that she was completely ignorant!

Specialization within the team

The issue of specialization is not very important when a study is conducted by a single person or a team of "experts," or if only one data collection technique is used. However, if more than one inexperienced field worker is going to be involved and multiple data collection methods are to be used, then specialization of team members for certain tasks is one strategy for obtaining high quality data within a short period of time. Specialization by task can be based on sex or on other criteria such as skills, abilities, and preferences of team workers for involvement in certain tasks. It is usually helpful for the team to meet during the day while its members are still in the community in order to cross-check information that they have obtained. This ensures that unusual facts or information that is contradictory can be further researched.

If a study utilizes more than one inexperienced field worker, the support, supervision, and motivation can be kept at a high level by approaching a community as a team, rather than by dispersing members in isolation in different communities over a longer period of time. However, once in the community, the team should divide up quickly to avoid a herd mentality. In a study in the Maldives, the project engineer (an insider) was used to establish credibility with officials and obtain data on water quality and quantity. Less experienced field workers were used for household interviews while a more experienced field worker conducted free-ranging key informant interviews and led group discussions. In a study in Indonesia, nine field workers were utilized: eight team members conducted household interviews; one specialized in conducting interviews using the local dialect; three women specialized in water collection; four men specialized in mapping. Additionally, two men and two women played a "health game" while others conducted key informant interviews using a checklist of issues.

A team of professionals

A quick study can also be done by a small team of specialists, each with specific responsibilities, visiting a community as a team. By spending a day in a community — visiting different neighborhoods; talking to a variety of people (men, women, children; leaders and ordinary people; rich and poor; those living near and those located at a distance from the roads) — a fairly detailed community picture can be drawn up.

In a World Bank study in Kenya, a team of six professionals — each specializing in two data collection methods — visited poor communities to better understand local perceptions of poverty and the access poor people had to water, primary education, and primary health facilities. Two people specialized in focus group discussions, while two facilitated mapping and wealth ranking; two others did household interviews and key informant interviews and talked with school children. The different methods were used to focus on many of the same issues. The team discussed its findings every night and followed up inconsistencies the following day. The teamwork enabled a detailed understanding of village realities by spending three days in each village examining many of the same issues through different methods.

Training field workers

In participatory research, the most important research tool is the field worker. The training of field workers, therefore, is of crucial importance. Even more important than training in specific data collection techniques is training in attitudes, behavior, motivation, and self-awareness. Unfortunately these aspects of training are often ignored, leading to data of questionable quality.

When field workers are told that their role is extremely important for the success of the study, they are being motivated to perform well. However, under difficult field conditions, when data collection efforts go on from early in the morning to late at night, realizing the importance of one's work to someone else's project may not be sufficient to revive flagging spirits and tired bodies. The human psyche is such that generally one is more apt to follow procedures and rules and do one's best when the rules are self-generated and when the project is one's own. The same is true for field workers involved in studies where close supervision is next to impossible.

Field workers must be convinced that they have a stake in the success of a study and that the study is their own endeavor This conviction can be created in two ways: by using participatory techniques in training them, and, if possible, using simple bonuses to encourage their participation. In addition to increased motivation and self-awareness, it is also important that field workers be thoroughly trained in ways of coping in the field and in administration of data collection techniques. (More detailed descriptions of techniques can be found in Section II.)

Raising motivation through participatory training

Since the basic principle of the participatory approach is to encourage and support people to develop answers to problems, training in participatory research should involve field workers in decisionmaking about the research process. Rather than giving them already developed instruments to administer (such as interview schedules, questionnaires, simulations), training should involve field workers in developing the instruments and in evolving the rules for their administration.

In this process, instrument development is combined with training as field workers help create the instruments, test and modify them, and explored under what circumstances the best results are obtained. As a result, they not only learn the process at a much deeper level but also develop a commitment to using the instruments in the most effective ways. This process, in turn, increases research reliability and validity.

In selecting candidates for training as field workers, prior experience as researchers or specialists in not necessary. In a study in Maldives, six field workers were chosen for a study on community and private toilets on the outer islands. None of them had previous research experience, and their education ranged from six years of primary school to some university education. In their training, the purpose of the study was first explained to them, setting an informal, non-threatening atmosphere and emphasizing equality in status. Issues and questions for an interview schedule were then presented for discussion. Items were added or dropped based on input from the participants; later, the interview schedule was pretested on a nearby island. Field workers determined how they would introduce themselves and how many interviews they would conduct. Field visits were followed by intensive, lively discussions of questions, answers, suggestions and ideas. After another round of trials, the schedule was finalized, as were the rules for administration and introduction. Despite hectic schedules, field workers remained highly self-aware of their own work and continuously assisted and reinforced each other's attempts. When one field worker had the misfortune of not finding people at home, another interviewer voluntarily helped her late at night, after finishing her own work.

A similar approach was also used in a study in Indonesia, where participatory training resulted in unusually high levels of commitment to the data collection process. It also led to the evolution of culturally appropriate techniques for interviewing, discussion groups, observational techniques of kitchen and water hygiene, and a health game utilizing pictures to tap people's concepts of "health" and "illness" of children.

Self-awareness

To obtain an insider's perspective during field work, it is essential, paradoxically, to maintain an outsider's perspective. This dual-function makes it crucial that field workers develop self-awareness and self-criticism to achieve involvement and detachment simultaneously in research situations. Anthropologists have the luxury of extended stays and gradual immersion into cultures which they are studying. Because field workers in quick studies cannot afford the time needed for gradual entry and learning, it is crucial that they enter the field with a certain degree of self-awareness, irrespective of the techniques of data collection they will use.

In a study in Indonesia, for example, field workers participated in training activities involving sentence completion, responded to pictures of village life, and played games in which success depended on cooperation and nonverbal communication. All resulted in bringing to the fore personality differences, values and beliefs which consciously or unconsciously influenced how the workers approached the field situation and their research tasks.

Participatory trainers integrate activities which clarify values and heighten awareness of one's assumptions about poor people, problems and capabilities as part of the overall training in participatory research.

The fit between data collection techniques and who collects data

If only two people are available for data collection, and this number cannot be changed, then the data collection techniques should be chosen to ensure that they can be implemented by two people.

In all other situations, the issue of who should collect data should be based on type of information, accuracy desired and methodology chosen. Constraints have to be kept in mind as do implications for sampling. It is useful to develop a summary matrix for the data collection strategy. On the following page is one such matrix developed in a study in which two field workers were involved.

The Fit Between Data Collection Techniques and Who Collects Data

Information Needs	Technique Chosen	Sample	Who Collects Data
Type, number of drinking water sources	Group discussions	Village chief and office staff	Male/female
	Observation	With ward/block leaders visit to 3 different types of sources	Male/female
	Mapping	After discussion groups	Male/female
Geographical terrain, rainfall, etc.	Observation	With ward/block leaders to visit 3 different types of sources	Male/female
	Group discussions	Village chiefs and office staff	Male/female
Settlement patterns	Mapping	After discussion groups	Male/female
	Group discussions	Village chief and office staff	Male/female
	Key informant	1 man, woman in satellite or distant settlement	
Distance of households to water	Observation, walking, mapping	Main water sources	Male/female
Functioning of sources	Interview, observation	Women living near major man-made sources piped system, wells, hydrants, etc.	Female
Reliability of water sources	Interview	Same as above, include any formal, informal caretakers	Female
Convenience of water sources	Interview, observation	Women living close to and a little far from sources	Female

127

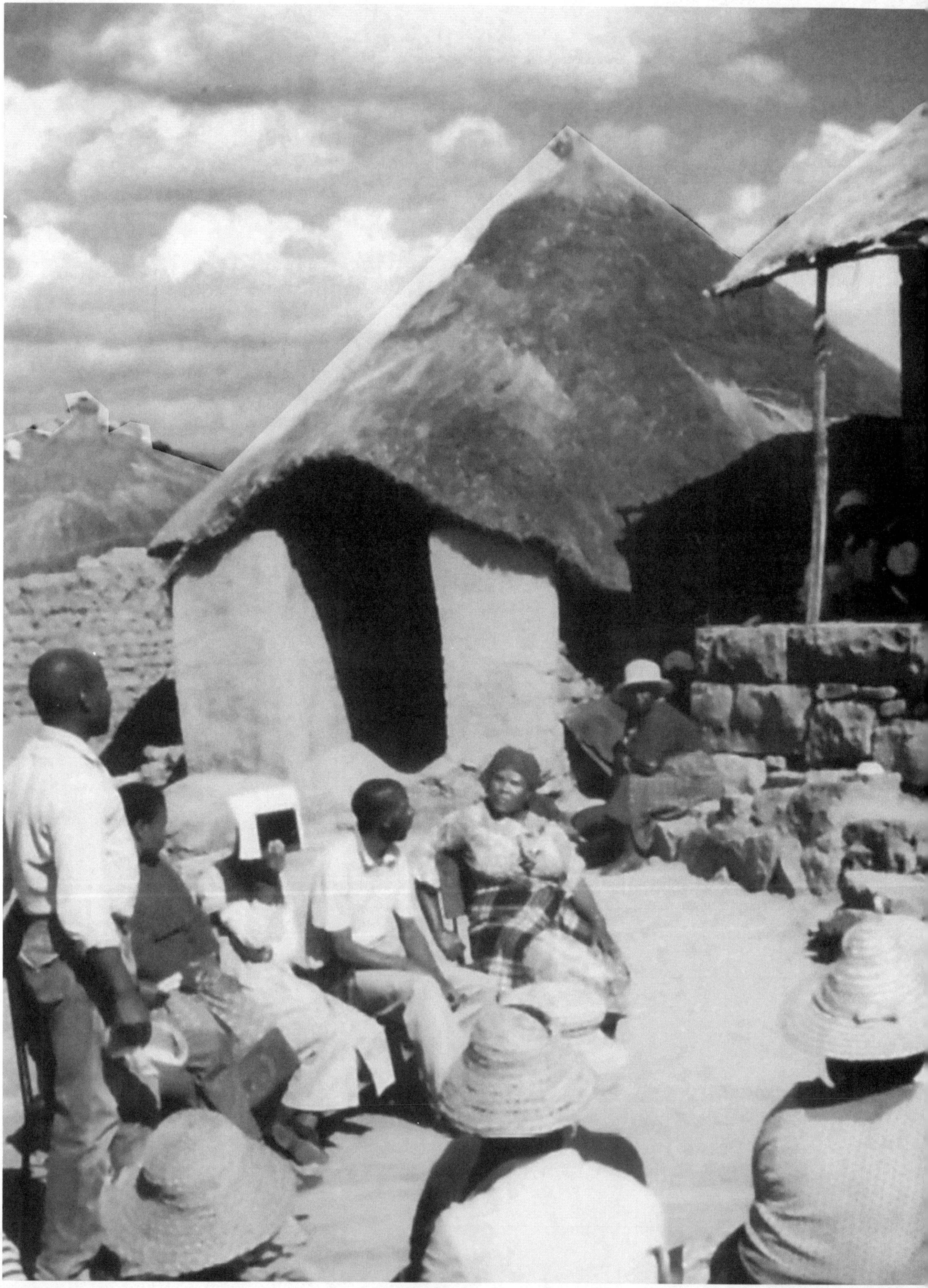

CHAPTER 9

DATA ANALYSIS, DISSEMINATION AND USE

Though data can be analyzed only after being gathered, planning for data analysis should be done during the early stages of planning the study. In participatory research it is often difficult to disentangle data analysis, dissemination and use since the same people are often involved in all three and the three components are occurring simultaneously. The following three examples clarify this process:

- In the Maldives, wood pieces were use to build two different miniature models of toilets which were presented to two community groups, separately to men and women. They discussed and analyzed the strengths and weaknesses of different features of the toilets — roofs, roofing material, placement of toilet seats, directionality, design of doors, orientation, and placement of skylights and windows. Men and women had lively discussions and disagreements about various features. For example, some men wanted low outer walls to prevent mischief makers from hiding and teasing women, whereas the women wanted higher walls so that they could not be observed from the outside.

  As different opinions emerged, men and women added, altered, or subtracted features and used the information on preferences to design the "preferred toilet." In fact they became so involved, that the next morning some men on the island had made detailed design drawings on paper with precise measurements. These were shared on other islands and also sent back to the Department of Water and Sanitation in Male, the capital city. Eventually, the recommendations were incorporated in the national five-year plan.

- In a project in Indonesia, data analysis was conducted during a day long workshop. The workshop brought together fifteen program staff, from village level workers and project managers, to community extension staff at provincial, regional, district, and sub-district level. The researchers began by presenting the findings without interpreting them; then, the raw data were turned over to the group for small group discussion. Since the data had been developed in villages with which the participants were familiar, they were not only analyzed and interpreted, but immediately used to alter extension strategies, set up a new schedule of training activities, and establish liaison with heads of the district authorities and the Department of Public Works.

Chapter Contents

This chapter explores the following issues:

- Who should analyze data?
- Techniques for data analysis
  - Qualitative data analysis
    - Narrative
    - Content analysis
    - Use of anecdotes
  - Quantitative data analysis
    - Tables, figures and graphs
    - Hand tabulation vs. computer analysis
- Dissemination and utilization of results
  - Characteristics of users
    - Primary interest and information needs
    - Time constraints
    - Language
    - Education and literacy
  - Media options
    - Seminars, discussion group and workshops
    - Slide-tape presentations
    - Tape recordings
    - Visual displays
    - Role plays, stories and games
    - Written reports
      - Content
      - Style and format
      - Length
      - What do users want?
- Follow-up action

At the same meeting, raw data were also presented through videotape footage of villagers conducting participatory self-evaluation activities with great enthusiasm. This was useful in grounding discussion in village reality and in overcoming extension workers' doubts about the capacity and energy of local water users groups.

- In a participatory research activity in a poor *kebelle* in Addis Ababa, Ethiopia, children and youth (aged nine to sixteen) participated in a week-long workshop which used a variety of interactive activities starting with needs assessment and extending to dissemination of findings. The group consisted of thirty girls and boys, most of whom were perceived by adults as troublemakers. Through self-analysis of their life situations, the group identified concrete problems and then chose one problem — lack of a safe recreation area — they particularly wanted resolved. Expecting assistance, the group then sat waiting for the facilitator to provide a recreation area! The following week the group called a special meeting and elected a sixteen-year-old youth, who had disdainfully sat apart during earlier sessions to deliver an appeal to the *kebelle* authorities. Their letter documented the children's analysis and argued for use of the *kebelle* recreation center at specific hours during the week. The findings were also used at another level by the external facilitator to convince project staff of the children's capacity to identify their problems and participate actively in seeking solutions.

Who should analyze data?

The issue of insiders or outsiders is equally applicable in the choice of who should analyze data and who should collect data. In many cases it may be necessary to use the same people who collect the information to analyze it. For example, if unstructured or semi-structured interviews have been conducted or group discussions held, then it is essential to have the person who collected the data do the analysis. This avoids problems such as what to do with information that has been gathered but not put on paper, or deciphering handwriting that is not easy to read.

It is particularly helpful to have the same people collect and analyze data when:

- Data are collected by methods other than pre-coded questionnaires and observation schedules;

- A purpose of the study is to build project capacity to conduct such research in the future;

- The research methods emphasize participatory procedures.

Techniques for data analysis

Options for data analysis and its presentation should be discussed and decided upon together with users. Obviously, data analysis techniques should meet the purposes and fit the nature of the data collected. The remainder of this chapter focuses on the many choices that are available.

Qualitative data analysis

All reports include some qualitative information that is of a descriptive nature. There are many data analysis techniques, the most common of which are narrative, content analysis, and anecdotes.

Narrative. As the term suggests, results are organized in a logical manner and written up in a narrative style.

Content analysis is a term used to analyze descriptive reports for trends, themes, or events. Content analysis can be utilized to summarize descriptive information or to transform qualitative information into quantitative information. Several anthropological software programs are now available to conduct computer-aided content analysis. These are useful if computers are taken out to the field so that field and research notes are typed directly into the computer. Otherwise, the process of first entering field notes can become cumbersome.

Content analysis is often used to set up the coding categories for quantitative tabulations. For example, if 100 people have been asked about advantages of using hand pumps, this information can be summarized through content analysis, coding and tabulation. The steps are as follows:

Step 1: Read every other answer and write down each distinct response.

Step 2: Pick the most frequent responses and state each briefly. Group each major response into one category.

For example: "Hand pumps are so much easier to use. I can send even my youngest child to collect water. The water is always clean and the pump has never broken down."

Categories: 1. Easy to use

2. Children can use

3. Clean water

4. Reliable sources - no breakdowns.

Step 3: Check each category to ensure it is mutually exclusive and that a coder will be clear which responses fall into which category.

Step 4: Complete the coding procedure.

Step 5: Tabulate frequencies for each response.

Use of anecdotes. Sometimes a few key anecdotes and quotes can effectively summarize the essence of what was said or concluded. Used in context, these can be very effective analytical tools. For example, during data collection for a process review in the Kwale region of Kenya, one key factor contributing toward an integrated approach to provision of water and sanitation services was the fact that the technicians and community development specialists had arrived at some understanding and respect for each other's work. The working relationship between the two — with the community development workers asking the engineers to slow down and wait for community readiness, and the engineers concerned with maximizing drilling efficiency — was graphically conveyed in the statement of the project hydrogeologist who said "We have finally learned that we must hurry the people slowly."

Quantitative data analysis

Quantitative data analysis is used to summarize information through percentages and other statistics. The relationship between two or more factors is explored by using such statistics and statistical tests of significance. Since a description of statistical tests can be found in any text of statistics and research methods, the emphasis in this section will be on pointing out primary options and their uses.

Tables, figures and graphs. Data can be analyzed using these methods based on the kind of data and the desired results. There are marked differences in people's ability to interpret even simple graphs, flow diagrams, and tables. These differences among users should be kept in mind in deciding on the use of different analytic techniques.

Hand tabulation vs. computer analysis. Decisions about whether to use hand or computer-assisted analysis of results should be made after consideration of all factors previously discussed, including the purpose of the study.

Advantages of hand tabulation:

1. Specialized, expensive, and sensitive machinery and skilled operators are not required.

2. Untrained people can be used.

3. The method can be replicated even in remote and isolated communities.

4. Hand tabulation is more amenable to user participation.

Disadvantages of hand tabulation:

1. Unless a systematic approach is adopted, errors are easy to make and difficult to track down in large data sets.

2. It is more difficult to do exploratory analysis that tests the association between different types of factors.

3. Hand tabulation sometimes does not generate credibility among decisionmakers.

Advantages of computer analysis:

1. Computer analysis often has more credibility among decisionmakers than do hand-tabulated results.

2. If the same type of data collection is to be institutionalized or repeated several times, the same format for data entry and the same program can be utilized.

3. Once data are correctly entered, exploratory analysis and comparative analysis over time are relatively easy to do.

Disadvantages of computer analysis:

1. The skill and specialized knowledge necessary to utilize computers properly may not be available, resulting in lengthy delays.

2. If computers are not readily available, it can create dependence on one or two overused and often distant sources.

3. In a participatory approach, computer analysis may perpetuate the notion of a "monopoly over knowledge" rather than increasing people's ability to generate knowledge and information systems.

4. Availability of a computer often results in overkill during data analysis, as it is relatively easy to think of all kinds of interesting and irrelevant or minor relationships.

5. Computer analysis is often more expensive than hand tabulation.

Dissemination and utilization of results

Written reports are only one way of sharing results. Before deciding on style, format, content, and medium of presentation of results, it is important to consider the fit between ways and means of disseminating results and the characteristics of different categories of users.

Characteristics of users

Primary interest and information needs. The interests of senior-level planners will be different from those of project personnel responsible for ordering hardware. Planners may be more interested in summary statistics, principles, and recommendations. Project personnel usually need more site-specific information. People within the community may be interested in an overview of findings and conclusions reached.

Time constraints. Very often, the people who most need the information have the least time to wade through thick documents. These time constraints need to be kept in mind while deciding on styles and formats.

Language. Suggesting that the language familiarity of users be considered may seem quite obvious, but fluency in major languages and familiarity with disciplinary jargon are all-too-often assumed. Too often, reports are written in a language that is foreign and little understood by those for whom the report is meant. Translation into the local language is a task that often is never done.

Education and literacy. Related to the language issue is the matter of differences in education and literacy. The kind of report written for a highly educated planner may be of little relevance to the community extension worker or to illiterate members of rural or urban low-income communities.

Media options

Written reports as the primary means of communicating results should not be a foregone conclusion. Other forms may be more important in attaining desired actions. Various options should be considered, keeping in mind the characteristics of users. Some of the options available are: seminars, discussion groups and workshops; role plays and games; video and slide-tape presentations; photo essays; visual displays; and flyers.

Seminars, discussion group and workshops have great advantages over written reports. In Sri Lanka, for example, a workshop was held with heads of relevant ministries soon after field work was completed and before formal analysis of the data had begun. This started a process of change eight months before the final report was officially submitted to the government.

Seminars can also be used to discuss preliminary findings, produce recommendations, increase credibility and heighten interest in and commitment to the study's recommendations. Perhaps the biggest advantage of such an approach is that the responsibility for doing most of the reading and for highlighting conclusions rests with the researcher rather than with managers and policymakers.

In a large water project in Asia, researchers who had submitted a very detailed, useful report had received no feedback from managers because the managers had not even read the document. An impending visit by project funders resulted in a half-day workshop when the researchers presented their findings using slides, anecdotes, and visual displays of large photographs. The impact of the workshop was so great that the manager proceeded to make major policy changes in the way the project was being implemented and also commissioned additional research for other similar projects.

In addition to the format, the tone researchers use in conveying their findings is equally important. In particular, findings should not be conveyed in a judgmental way. For example, during a planning seminar in an Asian country, researchers were asked to present study findings to thirty heads of ministries and departments as well as heads of districts. One key finding was negative: villagers who had previously contributed to building community water systems were no longer willing to do so. The reason: they no longer trusted local officials with their money because the water systems had, in essence, become private systems controlled by the wives of the chiefs and village officials.

Without linking the issues of money and control during their presentations, the researchers discussed people's willingness to raise resources to improve their lives. Examples were given and then a simple drawing was made and the researcher said, "One fascinating finding was that the piped systems were always located in such a way that the main stand post was in the yard of the village chief."

The officials in attendance laughed and during discussion linked this finding to people's lack of trust in village officials.

Slide-tape presentations to disseminate results can be an effective way of capturing interest of a wide range of audiences.

Managers of a PKK project in Indonesia did not write an end-of-project report but produced a simple eleven-minute slide show which has received wide dissemination in Indonesian villages, as well as in capital cities from Nairobi to Washington, D.C.

Video presentations have also been effectively used as a substitute for project documents and for fundraising purposes. Raw footage of community life, changes in the water situation and a project's impact on women can be powerfully portrayed on camera.

Tape recordings can be effective in capturing the spirit of dialogue, co-operation, and excitement about a project and its community-level activities. It is particularly effective in the hands of community members when they are asked to convey their own stories. Tape recordings of seven to nine-year-old children telling traditional stories, proverbs and riddles and describing herbal treatments for diarrhea in rural Zimbabwe were effective in convincing planners about the importance of involving children in decisionmaking.

Visual displays with or without workshops are an effective tool both at the community level and in agency headquarters. In Thailand, a community map marking all water sources was posted in the community center and quickly became a way of tracking community progress.

At a workshop, displays of research tools — community maps, pocket charts and photographs of large groups of participating men and women — helped convince planners that participatory tools to help community people in data collection activities were available, did work, and were worth investing in. Similarly, simple fliers and posters can be developed which highlight a few key findings.

Role plays, stories and games based on studies can be extremely important ways to make people aware of issues and results without appearing to be preaching. These media are especially important for transmitting information that is sensitive, and are useful in creating interest and setting the stage for follow-up action for training.

For example a children's board game called "snakes and ladders" can be modified to highlight issues that help or hinder community participation, or to show that the majority of water sources used by people are unsafe for drinking. In this game, two or more players roll dice and move pebbles or coins around the squares on a board. Some squares have a direction that aids participation; others involve obstacles such as: "village head installs taps near his house"; "money disappears"; "meeting held in the morning, so no women attend"; and "contractors work without informing village". (For more detailed information about this activity, see Section 2.)

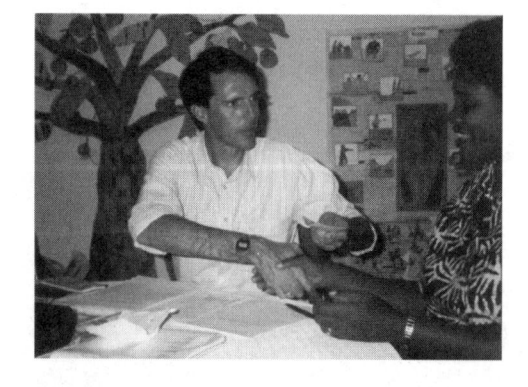

Similarly a role play, a puppet play, or story telling can make the process of disseminating results more interesting. These techniques help people pay attention, get involved, and remember. Participants become more likely to include the new information they have learned in their future decisionmaking.

Written reports. The purpose of highlighting alternatives to written reports is to show that these options do exist and can be effectively used by researchers. However, there is still an important role for written reports that are accessible, clear, and user friendly. Written information can be used by a wide array of people — community members, community organizers, district planners, regional planners, donor agencies, academics, and others. But it should be emphasized that even the best researchers do themselves a disservice by writing reports that do not create interest and are not read by those who need the information the most.

Not everyone is equally interested in all the details of the study. A policymaker is unlikely to be interested in eight pages of detailed socioeconomic background, including what percentage of people own chickens. A planner might be interested in a summary statement of how socioeconomic status was measured and what people could afford to pay for improved water and sanitation if they were interested.

Content. It is important to distinguish in some form between facts, interpretation, conclusions, and recommendations. Depending on the media chosen, content can be in the form of words, figures, table, pictures, or artifacts (such as maps and games).

Style and format should be considered carefully to ensure that users of results are presented with information in a way they find easy to follow and that catches their interest.

Length. Even if a report, game, or seminar is appealing, if it takes too long people may lose interest before the end or may not be available for the entire time.

What do users want? During the planning stage of the study, it is wise to talk to the users about their expectations and preferences regarding presentation, style, format, length, and medium. Various alternatives can be suggested and discussed. Once results have been analyzed it is usually worthwhile to involve users again as much as possible or at least in the key decisions.

Follow-up action

Traditionally, researchers have not considered themselves responsible for follow-up action. This assumption perhaps has been based on a strict and somewhat artificial division of labor between researchers and the users of the findings. This is characteristic of research approaches in academic settings which stress the objectivity of scientific research. The attitude has characteristically been "I am a researcher. I am responsible for the research study. I will do my best and then hand over the results to planners. Then it is up to them to use it or not."

This attitude must be discarded by those involved in participatory research. Development practitioners, whether using the conventional or more participatory approaches, must be more concerned with the utility of their study efforts. In the final analysis, the purpose of the study is achieved only when the process of doing the study and its results together further the goal of improved water and sanitation.

SECTION TWO

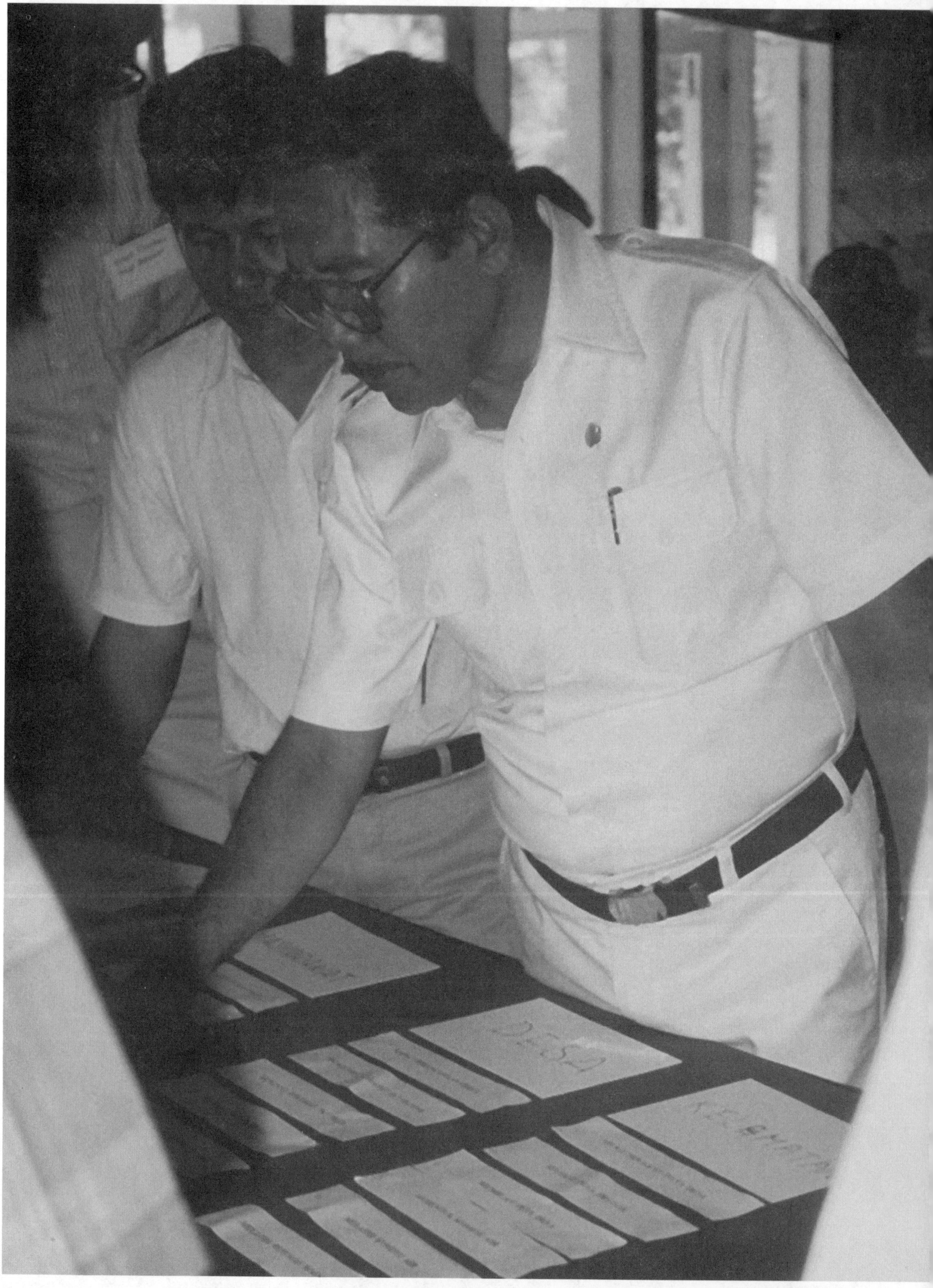

Chapter 10

Participatory Research Activities

Introduction

This chapter contains a wide range of participatory activities for use at both the community and agency levels. In selecting which research tools to use in planning community-based water and sanitation projects, an essential guiding principle is shared decisionmaking and local empowerment. The central objective is vesting control with the community or agency staff. This principle must in fact govern the entire participatory research strategy. The activities and tools utilized then become, in and of themselves, instruments for community level and agency staff capacity-building.

Special care must therefore be taken to make the research experience a user-friendly one; innovative enough to arouse curiosity; one that evokes participation and makes it pleasurable and useful; a process that dissolves hierarchical barriers between researchers and local people, opening up free communication among all actors.

Designing materials and activities that fulfill such sensitive functions requires skill and an unwavering commitment to the goal of local empowerment.

The benefits of using participatory tools

Within this section are examples of ways local people can be progressively involved in the research process without being overwhelmed by it. These tools are designed in a way to empower and enable local people to experience a new sense of human dignity, competency and ownership.

The majority of activities and tools described in this chapter engage people in investigating their own situation and circumstances, in comparing and assessing the data generated through this process, in selecting and prioritizing it and gaining a new perspective on everyday problems, their causes, effects, interconnections, and possible solutions. Cumulatively, these tools build local capacity for decisionmaking at increasing levels of complexity.

The benefits to the researcher in employing these activities and tools are substantial. With a steady increase in community and staff self-assurance and analytic skills, the field worker gains access (often for the first time) to a wealth of facts, perspectives and insights from the community's and the staff's own store of wisdom and experience. This type of sharing, which proceeds from mutual trust, is facilitated by the informal and egalitarian relationships in which participatory transactions are conducted. This degree of openness is unlikely to be achieved when research is conducted in a hierarchical style, separating out the experts from the respondents and circumscribing their roles.

At the same time, the insights gained through the use of participatory tools can also have pitfalls. Researchers may be tempted to exploit these tools primarily for the extractive purpose of gathering data for planning. Instead of contributing to capacity building, this experience tends to leave the participants puzzled; they see it as a pointless series of games which have been thrust on them and seem to lead nowhere. This produces frustration and creates a sense of being manipulated and used.

Capacity building requires much more than exposure of participants to a set of participatory investigative activities. It is the result of a sustained process involving new experiences, reflection, analysis, exploration, decisionmaking, acting and evaluation. At some point in this process, the researcher's role must give way to the facilitator's role and the human development objective must override the more extractive data gathering objective.

This requires that field staff have skills in facilitation which go beyond the skills normally available in the researcher's training. The implications are obvious, both for training research staff and for teaming up with other staff specialized in nonformal education and facilitation. When all staff share a vision of community self-direction and utilize participatory activities primarily for their enabling and empowering goals, everyone stands to gain.

Sensitivity to gender issues

Gender issues are deeply entrenched and often emotionally laden. Focusing on gender issues often involves challenging deeply entrenched values and time-sanctioned practices of the majority of male members of communities and agencies. This can even be true of women who have internalized negative values and images about themselves. In these situations, participatory investigative tools can be of critical value.

For example, simple methodological tools can be used to enable women to analyze their life. This examination can begin with a simple issue such as how they perceive their priority problems. If water is a perceived problem, the women can analyze what water sources they use, how many trips they make to the water source, what else they could do with their time, what roles they play vis-a-vis men at the level of the family, the farm or the community, and how they feel about different technology options. Similarly, both men and women can be involved in gender diagnosis through activities that focus on division of labor, sources of household income, relative control over resources, and roles played in decisionmaking at different levels, from personal to community level.

However, here, too, using a variety of diagnostic activities and tools in a mechanical sequence in a straightforward research stage, can be self-defeating. Opportunities for deeper reflection on sensitive issues are lost. Changes in gender-biased attitudes are unlikely to result unless processing time is allowed and importance attached to fully examining, questioning and evaluating the outcomes of each activity through consultation, problem-posing, debate and mediation, thus preparing the ground for a larger process of societal examination. The gender-focused activities contained in this section are therefore intended as a critical starting point for such a process.

Agency level participation

Institutional mechanisms in support of community participation pose a different problem, particularly at the project management level where the authority to make policy and program decisions is vested. Here the central objective is also empowerment, but in a different sense: not empowerment to increase authority and control, but rather to be able to increase the capacity of others to have a greater say in management. All project staff who previously functioned on the basis of issuing directives, handing down plans and requiring compliance to pre-set targets will now need to change to a more open and flexible approach. They will need skills by which to stimulate and assist community members playing a larger and more critical role in community decisionmaking. Project managers themselves will need to scrutinize their own policies or practices to ensure consistency in applying participatory principles throughout the management system.

Activities and tools are therefore needed to assess where project management stands conceptually in regard to community participation, what they understand by it and how they perceive it can be translated in action. Theoretical endorsement is not enough; there must be a high level of commitment to make the concept work in operational terms through appropriate budgetary provision, programming flexibility, provision of trained staff and personnel policies that prioritize and reward initiatives to promote community level institutional capabilities.

Participatory research tools applied at this level have, however, to be adapted, keeping in mind that senior personnel are generally too busy with administrative matters to spare time for research purposes. The design of investigative activities at the agency level must also take into account the degree of willingness of staff to experience these tools by participating in them directly. The informality of participatory research may be disconcerting to those who are particularly status conscious. Special care is therefore needed to ensure that top management feels that participatory research tools are both appropriate and challenging enough to avoid their dismissing the whole endeavor as frivolous child's play.

Chapter Contents

This chapter contains descriptions of thirty-three participatory activities. In most cases, each activity write-up provides the following information:

- Purpose
- Background
- Time required
- Audience
- Materials needed
- How to conduct the exercise

As appropriate, the descriptions also suggest adaptations of the primary method for conducting the exercise, as well as the source of the activity when known. The activities are divided into the following categories:

A. Community Profiles

1. Map Making
2. Community Self-Portraits
3. Oral History
4. The Demographic Pebble Activity
5. What is Poverty? Who is Poor?
6. Wealth Ranking
7. Rating Scales
8. Pocket Chart-Water Use

B. Gender Analysis

9. Control and Access to Resources by Gender
10. Task Analysis by Gender
11. Women's Lives: A Needs Assessment
12. Women's Confidence
13. Women's Time Management
14. Women's Involvement in Operation and Management (O&M) of Water Systems

C. Semi-Projective Techniques

15. Incomplete Sentences and Open-Ended Stories
16. Creating Open-Ended Stories
17. Role Playing Open-Ended Stories

D. Games and Simulations

18. Health-Card Sorting Game
19. Health Education Game
20. Bintang Anda Game
21. Concepts of Health and Sickness: A Health Game
22. Open-Ended Snakes and Ladders
23. Health Seeking Behaviour

E. Technology-Related Activities

24. Local Criteria for Choosing Among Technology Options
25. Pump Repair Issues
26. Understanding Pipe Water System Through a Model

F. Management Tools and Problem Identification

27. The Problem Tree
28. The Bottomless Pit Exercise
29. Understanding the Decisionmaking Process
30. Expectations of Community Participation
31. What is Community Management?
32. Evaluating a "Unified Vision"
    - Assessing Goals and Objectives
    - Role Perceptions
33. The "Mess" Exercise

A. Community Profiles

Community profiles can be used to help start a dialogue about collecting data on the sociological and economic characteristics of a community. By creating community profiles, participants gather important basic data that can then feed into follow-up planning activities.

The following activities are found in this section:

1. Map Making

2. Community Self-Portraits

3. Oral History

4. The Demographic Pebble Activity

5. What is Poverty? Who is Poor?

6. Wealth Ranking

7. Rating Scales

8. Pocket Chart Water Use

1. Map Making

Purpose

To gather information about a community by enabling community members, including children, to represent their community and its issues through drawings or three dimensional models.

Background

Map making can be used to identify many kinds of information, including:

- locating the major resources of a community and its physical features (rivers, ponds, markets, roads, health posts, schools, etc.);

- identifying existing communal and private water and sanitation facilities, and which are used by men and which by women;

- focusing on health hazards and resources that promote health;

- providing a visual picture of the community social structure (ward and kinship groups and boundaries, rich and poor, male and female local leaders, etc.)

Maps can be of any size. They can be drawn on paper and superimposed on one another to show change or to add new features; they can also be made on the floor or on walls in schools, people's homes, at the community center or the village office. Maps drawn in public places are an extremely effective community monitoring device.

Time

1/2 - 2 hours

Audience

Primarily community members; also useful for agency trainers, project staff and field workers.

Materials needed

Materials needed vary, depending on whether the map is a drawing or composed of three-dimensional models.

Maps can be drawn on the ground with a stick, or, if the group wants to produce maps which can be used in future discussions, large sheets of paper with felt pens or other drawing materials can be used.

Three-dimensional maps can also be built using wood or scrap material. String and ribbons can be used to show connections between activities, geographic areas or socio-economic groups. Twigs, leaves, seeds, stones, wire, clay, shells, or other found objects can be used to symbolize groups or facilities with a community.

How to conduct the exercise

1. Begin by determining whether the mapping will be carried out within one group or several. If the group is large, participants may want to divide themselves into smaller sub groups to depict different parts of the village which can then be joined together into a larger map.

   In most contexts, it is important to have men and women map separately because there are usually strong gender differences in perceptions about the community.

   It is also important to be aware that different people have different perceptions. Whenever mapping activities are conducted, sufficient time should be allowed to resolve differences and reach consensus. People whose perceptions differ dramatically should make separate maps. These differences will reveal important information through discussion.

2. Provide participants with mapping materials and ask the group to use them to describe their community by making a map of it. Remind them that they are the "experts" in terms of knowing their own community life. Reassure them that they are free to plan and produce the map in any way they wish.

3. After introducing the activity in this way, the researcher or facilitator conducting the session should "let go", letting the participants take over the process.

4. Ask some members of the group to use the map to take the other participants on a tour of the community, including the topography, demographics, aspects of the lives of the people, those things that people are proud of, and those they see as problems.

5. Based on the map and how people have described their community, initiate discussion on any of a range of issues: water supply and sanitation needs; particular concerns of women; community history (what the community was like in the past; how it has changed; what made it change); community aspirations for the future (in what ways can the community be improved; what aspects of life need to be changed; how can that be done?);

Adaptations

When to carry out mapping

Maps can be drawn at the beginning of projects to establish baseline data, during implementation for monitoring, and over time to evaluate progress or change.

Evaluating design criteria

Map overlays can be used to evaluate whether a proposed piped system and its distribution network will meet design criteria and reach the target population. When using maps for this purpose, field workers, together with local people, can map the community settlement pattern. In this case, The maps should be the same scale as maps of surveyors. The proposed piped system is then overlaid on the new map with details of house locations. Using the design criterion of number of people per standpost or maximum distance of household to standpost will reveal whether the proposed piped system will attain total coverage. Thus, maps can be effective in persuading engineers to change the placement of secondary and tertiary distribution lines and the location of standpoints.

Map making with children

Small maps made by groups of children divided by location can be pieced together to make a community map.

Map making by researchers and planners

Map making of communities by researchers can be used for training. Researchers are asked to build a community map based on their recollections or perceptions of a rural or urban poor community. Planners in rural or urban areas can use mapping for targeting facilities for the poor, designing key garbage collection points or routes, and a variety of other purposes to rationalize allocation of resources or to compare their knowledge about a particular issue with that of local people.

2. Community Self-Portraits

Purpose

To generate dialogue on specific community issues and problems through interaction with visual images, reflection and discussion.

Background

Because it is necessary to first describe something in order to identify critical issues or problems associated with it, accurate description of a community is an important part of the research process. The medium of drawing as a descriptive technique is particularly suited to poor rural or urban communities. Even a simple, small drawing can be a tool that sheds light on a community's circumstances and generates dialogue and action.

It is often most effective to create a community self portrait in association with some other important local event, such as a community celebration or meeting.

It should be noted that drawing is only one way for a community to portray itself. Some other ways include socio-drama, songs, storytelling, artifacts, photography, and audio and videotapes.

Time

1 - 2 hours

Audience

Community members.

Materials needed

Paper, crayons and tape.

How to conduct the exercise

1. Begin by affixing a large sheet of paper to a wall. Ask participants to verbally describe aspects of their community: what do people do?; where do they live?; and so forth. Once people have called out their responses, ask them to drawing their descriptions on the paper.

2. Encourage community member to make the portrait as detailed as possible — for example, distinguish rich homes from poorer homes, and depict specific activities by gender (water gathering, taking care of the children, planting and harvesting). Relationships and systems which are normally hidden can be diagrammed. For example, people often realize the underplay between economic and political systems. Sometimes as people are drawing, a shift in focus occurs from describing everyday experiences to questioning why and how specific problems exist. Naming these issues and problems is a small, yet critical, aspect of a whole research process.

Source

Adapted from *Drawing from action for action: drawing and discussion as a popular research tool.* 1981. Participatory Research Group. Working Paper 6, Dean Marino. Toronto, Canada.

3. Oral History

Purpose

To understand the community's history, attitude and values toward its leaders, women, production systems and government action.

Background

In short cut, rapid research, there is always a danger of taking an ahistorical perspective by not understanding the context of a situation, the degree or lack of change or the reasons for current problems. To increase the depth of understanding and achieve a perspective on a situation over time, it is important to get a sense of the history of a community, what people's circumstances have historically been, and what their goals and aspirations are.

As the term implies, "oral history" is the verbal history of people's situations. Because oral history is not written down, not only are the words people use important, but so too are how they are used, and the tone and emotion with which they are spoken. In eliciting oral histories, the person asking the questions becomes part of the process by recording the words and by interpreting their hidden meaning.

Oral history can be of places, events or lives. Life stories are extremely important in understanding the concerns, problems, critical incidents, and changes that occur through an entire life cycle. In telling their stories, people reconstruct the past as they remember it. The subjectivity of the story is its revealing feature. What people tell or don't tell are keys to interpreting how they feel about certain situations.

Oral histories can be particularly useful in water and sanitation projects to understand the following:

Village history, its geography, ethnic composition, when and why it was formed, who were its leaders, and what changes took place over time. In recording village oral history, older men, women, and traditional leaders make good key informants.

Life history of women, the critical events of their lives, their power and status in the community. Both older and younger women should be interviewed in order to understand generational differences in points of view.

History of water sources can reveal people's attitudes toward water and its sources, the need for water, understanding water pollution and conservation, water divination, and explanation of failed water systems.

Time

1 - 3 hours

Audience

Community members.

Materials needed

Extensive notes can be taken; alternatively, tape recorders can be used.

How to conduct the exercise

1. Begin by setting the right atmosphere. People are usually more relaxed in the late evening or at night once the day's work is over. People are more likely to talk if they believe that the person asking the questions is genuinely interested and non-judgmental. Women should be interviewed in privacy if their life histories are being recorded.

2. Conduct a conversational interview, keeping the main purpose in mind.

3. If tape recorders are used, be aware that the transcription process can take quite a long time.

Adaptation

The oral history of projects and agencies can provide valuable insights about why it is difficult to induce change, the role of political patronage, and the organizational culture or leadership of the agency.

4. The Demographic Pebble Activity

Purpose

To involve community members in conducting a mini-census to reflect on the water, sanitation, and health care needs of families of different sizes.

Background

Community members often do not understand the functions or trust the motives of investigators who come with paper and pencil in hand, ask questions, make notes and then go away with information without having produced any apparent benefit to the community.

This activity is one of many ways to involve community people in conducting their own rudimentary needs assessment. Through this kind of direct experience, they can develop the understanding and skills needed for more complex survey processes.

Time

1 - 2 hours

Audience

Community members.

Materials needed

Stones, Color-coded pebbles, beans of different sizes, or coded matchsticks or sticks of different lengths, in sufficient quantity to represent ten to twelve large families. Pebbles of different sizes can represent different gender and age groups.

Containers labelled with pictures of family members differentiated by gender and age (elders, father and mothers, young men and women, teenagers, boys and girls, children under age five).

How to conduct the exercise

For the purposes of the following example, pebbles will be used.

1. Organize the participants into a group not larger than ten people. If there are more than ten participants, divide into smaller groups.

2. Place the pebbles representing the different family members inside the appropriate containers.

3. Ask the participants to select the number of pebbles from each container that represent the composition of their family. *Note: through special color coding, size differentiation or, in the case of matchsticks, "dressing up", humor can be introduced in the activity to make it more enjoyable.*

4. When everyone has picked materials from the cans to represent their "family," remove the extra materials from the container but leave the containers and their labels in place. Also, leave a sample pebble, bean or matchstick next to label to remind the participants about the coding system.

5. Ask individuals in the group to each take turns putting their "family members" back in the containers according to the designated categories; then have them count the number of pebbles in each container and develop a composite profile of the families in the group.

6. Invite reflection on the total number and what specific needs each of those categories of people represents for the family.

Discussion can move from general issues (such as whether unemployed youth or school dropouts in the village represent a problem) to sector-specific needs (for example, which group is most affected by the lack of access to safe drinking water).

Adaptation

Instead of using containers, a matrix can be drawn on the ground; the boxes of the matrix can represent different family members.

Agency staff can also do the exercise to conduct a census of staff profiles by age categories, gender and training.

Source

Lyra Srinivasan. 1992. *Options for Educators.*

5. What is Poverty? Who is Poor?

Purpose

To elicit a community's perspective on what poverty means to that particular community; and to reach community consensus on who is poor and hence qualifies for targeted program assistance.

Background

This activity can be made simple or complex; if the researcher wants the participants to categorize individual community families by wealth-ranking, it is necessary to have available a register with the names of all households.

This activity is best conducted in small groups of 8 - 15 people.

Time

1 - 2 hours

Audience

Primarily community members; also useful for trainers, project staff and field workers.

Materials needed

Three drawings of a man, a woman, and a couple.

Three labels: rich, average, and poor.

Smaller cards depicting pictures of objects, possessions or types of people commonly found in the community and that are likely to reflect differences in economic status among families, such as:

radio	horse	television
bedroom furniture	trees	bicycle
garden	crops	maize
chickens	fruit tree	money
cart	vegetables	goat
bonded laborer	village money lender	

Some small, blank cards participants can use to draw any possessions or objects that may be missing from the above set.

How to conduct the exercise

1. Place the three labels — rich, average, and poor — on the ground side by side. Then spread out the cards randomly below the labels so that they can be viewed by all of the participants.

2. Ask the participants to discuss and categorize the cards by placing them in vertical columns under the three labels, depending on whether the possessions are likely to be owned by rich, average, or poor people in the community. Ask the participants to draw or write on blank cards any materials or attributes (such as powerlessness, happiness, sense of belonging, children) that are missing from the small cards.

3. After a consensus has been reached on who is likely to own which possessions, ask participants to identify three of the small picture cards that best characterize each group.

4. Remove the labels and again mix up the cards. Now use the large drawings of a man and a woman to indicate whether a household is headed by a man or woman. Ask participants to categorize the cards again on the basis of whether there are differences in wealth and well-being between male and female-headed households.

5. Finally, ask the participants to categorize actual families in the community in terms of whether they are rich, average, or poor. Before beginning, ask the participants whether this process should be done with confidentiality. Names of families can be written on slips of paper. Allow for ample discussion until consensus is reached.

6. Based on the categorization the group has carried out, ask the participants to discuss what the priority needs of each group are.

Note: *Sample materials for conducting this activity can be found in the Participatory Development Tool Kit.*

6. Wealth Ranking

Purpose

To involve local people in ranking the relative wealth and poverty of families in the community based on their own criteria, for purposes of better targeting development assistance.

To enable staff to understand how community members perceive differences in poverty and wealth.

Background

This activity is based on the experience of an Oxfam team in Sudan while conducting a rapid nutritional assessment in 1988–89 to measure the rates of malnutrition in drought stricken communities. This approach helped the team better understand people's perceptions of differences in wealth among families with malnourished children; information gathered facilitated the targeting of families that were most in need of food.

This activity can assist in providing information that leads to more effective targeting of assistance in many areas that affect a community's quality of life, such as water and sanitation, health and basic education.

Time

1 hour

Audience

Community members and project staff.

Materials needed

Blank cards and writing implements.

How to conduct the exercise

1. Ask the village development committee, or other appropriate local organization, to identify three members of the community as volunteers for this activity.

2. Discuss with the volunteers their perceptions of the differences between rich and poor families in the community. Allowing for the differences in views the volunteers may have, arrive at some generally agreed-upon definitions or phrases to use for the purposes of categorization of families by level of wealth.

3. Ask the participants to identify specific households that will be ranked through this activity, as well as potential recipients of aid. Each volunteer should write the name of the head of each household on a separate card.

4. Have the three volunteers sort out their heads-of-household cards into groups according to their perceptions of differences in wealth. If any group contains more than 40 percent of all families it should be further subdivided. The researcher should have the volunteers make their decisions independent of one another, so that their rankings can be cross-checked at the end of the activity.

5. When the households have been categorized, ask the participants to discuss the characteristics of each group. Explore whether the wealth rankings given to the selected families is also generally applicable to the families that were not chosen for ranking.

Source

Helen Young. 1990. *Rural Evaluations*, RRA Notes 8. London: International Institute for Environment and Development.

7. Rating Scales

Purpose

To measure how a group, a committee, a project or a leader is functioning.

Background

In most cultures, stars have positive connotations. Accordingly, stars of different sizes can be effectively used as a rating scale in a variety of ways. Depending on the context, different size stars can represent values such as excellent, average, poor; or very good, good, poor; or very important, important, not important; or many, few, none.

Pictures representing an issue or situation can be categorized by placing them under different stars; ratings given by different groups or the same group for different activities can be serve as a focus of discussion. Scales are most likely to be appropriate if developed in discussion with local people.

Whenever a rating activity is conducted, it is very important to involve women in the task in groups separate from men, at least initially.

Time

1 - 2 hours

Audience

Community members.

Materials needed

For this activity stars of three sizes are used. The largest star should be twice as big as the medium-size star; the small star should be half the size of the medium star.

How to conduct the exercise

There are many variations on the Rating Scale exercise. Here, for example, is how the activity can be used to enable water user's groups to rate their overall performance and to evaluate the contribution of key people and activities to the functioning of the group.

1. Place each of the three stars on the ground in descending order of size. Explain to the group that, depending on the context, the stars represent excellent, average, poor; or very important, important, not important.

2. Then, display and explain cards depicting water group functions and key people associated with water groups. The cards can depict, for example, group cooperation; a village leader; sanctions; fee collection; angry group members; planning and design; an extension worker. First, ask the participants to discuss the effectiveness of each person or activity in their own water group.

3. Then, ask the participants to place each of the pictures underneath the appropriate size star to rate how effectively it functions.

4. Once consensus is reached, ask the group to explain and discuss its ratings.

5. Ask the group to give *itself* an overall rating.

6. Encourage the discussion to focus on follow-up planning to take corrective action to increase the effectiveness of various people or processes of importance to the functioning of the group.

Note: Sample materials depicting water group functions and key people associated with water groups referred to above can be found in the "Participatory Development Tool Kit."

Adaptations

- *Five Point Scale.* Measures of various perceptions and attitudes toward factors affecting the quality of life can be developed using a five point scale: very satisfied, satisfied, neutral, dissatisfied and very dissatisfied.

- *Symbolic Scales.* Commonly accepted symbols within one culture may not be interpreted in the same way within another culture or environment. It may be appropriate, therefore, for people in a community to suggest symbols representing a situation or quality. For example, line drawings of happy and sad faces are not understood in many rural areas. Hence, use symbols with caution and, if possible, have local people develop them.

- *Pictorial Scales.* Pictorial scales are those which use pictures to indicate the relative values along a graded scale. Pictorial scales are the most concrete and easiest for people to understand.

8. Pocket Chart—Water Use

Purpose

To help community members learn a new way to assess and analyze their situation.

Time

1 hour

Audience

Primarily community members; also useful for trainers, project staff and field workers.

Materials needed

A pocket chart consists of rows of paper or cloth pockets, usually four to six horizontally and six to ten vertically. A set of pictures is attached above the top row of pockets. These pictures represent areas in which data are needed, such as different sources of domestic water supply. Each of these pictures is placed at the head of a vertical column. If desired, pictures can also be attached down the left-hand side to indicate other variables, such as how the different water sources are put to use by community members. However, in order to avoid confusion, the facilitators should use only one variable on a column at a time.

Materials required to carry out this exercise include:

- three to ten picture cards;
- a pocket chart (either as described above; using cans or pots set on the ground beneath the pictures; or drawing a matrix on the ground);
- paper slips, leaves or seeds for voting.

How to conduct the exercise

1. Begin by creating a pocket chart as follows:

   Across the top of the chart, place three of the cards depicting water sources (a river, a well, a pump, and so forth).

Down the left-hand side of the chart, place three or more cards showing different water uses (cooking, washing, drinking, and so forth). The number of cards used is based on the detail and level of complexity desired.

Place small cloth or paper pockets or other containers along each of the rows you have created.

2. Place the pocket chart in a location that is accessible, but also where voting can be done confidentially.

3. Explain to the group that the pocket chart can be used to determine how different sources of water are used by the community.

4. Illustrate how the balloting is performed by placing a slip of paper into a pocket or container to indicate a choice or preference. Remove the slip after the demonstration.

5. Ask for six volunteers to perform the voting. Give them enough ballots so that they can vote on how they use each source of water. Have the participants vote one at a time. If the entire group wants to vote, organize the voting accordingly.

6. When the voting is complete, ask another group of volunteers to remove the voting forms and tabulate the results.

7. Discuss the patterns of use that emerge and record the findings. Engage the group in a discussion about the meaning of these findings. For example, "Why do so many (or so few) people prefer one source of water for washing over another? Is this sample representative of most people in the village? Do preferences have any effect on health and well-being? Are there seasonal differences?"

9. Once the activity is understood, it should be taken over by the community and used to assess and analyze information about other issues they face.

Note: *Sample materials for conducting this activity can be found in the Participatory Development Tool Kit. The pocket chart method was developed by Lyra Srinivasan.*

B. Gender Analysis

Gender analysis focuses on understanding and documenting the differences in gender roles, activities, and needs, in a given context. By disaggregating qualitative and quantitative data by gender, gender analysis highlights the special roles and learned behavior of men and women based on gender, rather than biological, attributes.

Gender analysis activities should be used to enable project designers to determine where to pay particular attention to the involvement of women. These activities can also open the minds of community members to the situation of women and where improvements can be made.

The following activities are found in this section:

9. Control and Access to Resources by Gender

10. Task Analysis by Gender

11. Women's Lives: A Needs Assessment

12. Women's Confidence

13. Women's Time Management

14. Women's Involvement in Operation and Management (O&M) of Water Systems

9. Control and Access to Resources by Gender

Purpose

To collect information, raise awareness, and enable the community to understand how access to and control of domestic and community resources varies according to gender.

Background

Gender analysis concepts tend to be abstract and controversial since they affect everyone. Visual tools have been found to be very effective in getting people, both men and women, to focus on gender issues without being threatened. In most countries where these tools have been used, they have been so successful in stimulating discussion that the process could go on for hours if the facilitator did not end it.

Time

1 hour

Audience

For community members (men and women); also extremely useful for agency staff, trainers, and field workers.

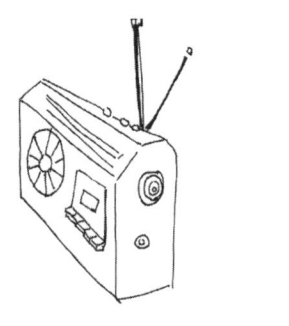

Materials needed

Large drawings of a man, a woman, and a couple.

At least fifteen cards depicting different resources and possessions owned by local community members, such as:

- cattle	- chickens	- bicycle
- currency	- trees	- vegetables
- furniture	- huts	- plants
- fruit	- pipe	- jewelry
- bags of maize	- donkeys	- horse/cart
- radio	- sheep	

Blank cards and writing implements.

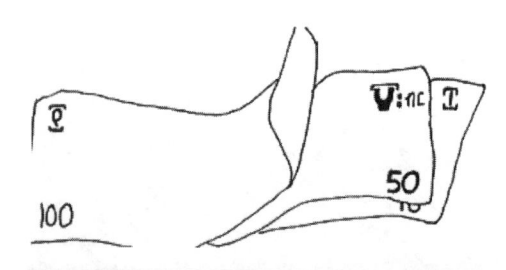

How to conduct the exercise

1. Place the three large drawings on a table or on the ground, in a row. Underneath these drawings, scatter the smaller cards at random. Include some blank cards.

2. Ask the participants to sort the cards by categorizing them under the three large drawings in columns, depending on who owns or controls the resources. If important resources or possessions are missing from the cards, ask participants to draw them.

3. Facilitate the discussion among the participants about why they made the choices they did. Make sure women are included in the discussion, or organize the exercise into two separate activities, one for men and one for women, and let them share the results of their activities.

4. Next, ask participants to focus on women's access to resources, even if they are controlled by men. Give people colored stickers to mark the resources that men own, but women can use. Focus the discussion around the use of the resources.

5. Ask participants to discuss what happens to women's access and control over resources in cases of divorce or separation. Ask participants to move the cards or introduce another color sticker to mark the resources that move out of women's control due to divorce or separation.

Note: *Sample materials for conducting this activity can be found in the Participatory Development Tool Kit.*

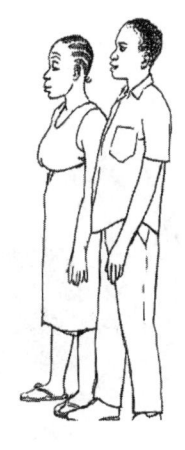

10. Task Analysis by Gender

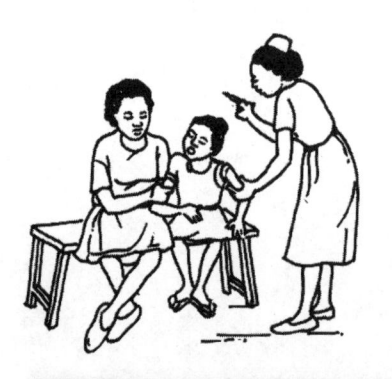

Purpose

To raise awareness within the community of how domestic and community tasks are distributed according to gender; and

To help the researcher understand the degree to which role flexibility by gender is associated with different tasks.

Background

Like the previous activity, a visually conducted task analysis is very effective in raising gender awareness and getting information on gender specific tasks in a particular cultural environment. Combined with resource analysis, it usually highlights the imbalance between women's access to resources and their workloads.

Time

1 hour

Audience

Primarily community members (men and women); also very useful for trainers, project staff and field workers.

Materials needed

Three large drawings of a man, a woman and a couple.

At least a dozen cards depicting daily household and community tasks being performed (such as growing crops, basket weaving, looking after a child). The pictures can be drawn using either male or female figures, regardless of whether it is a man or a woman who usually performs the task in question. Blank cards should also be provided so that participants can draw tasks not already included in the set.

How to conduct the exercise

1. Place the three large drawings on the ground, in a row. Underneath these drawings, scatter the smaller cards at random.

2. Ask the participants to sort the cards by categorizing them under the three large drawings in columns, according to whether the task is generally performed by a man, a woman or both.

3. Let the participants take over the exercise and conduct the discussion themselves.

4. When some degree of consensus is reached, initiate a discussion among the participants about why they made the choices they did. Be particularly sensitive to including women in the discussion.

5. Ask the group to analyze the workloads — both the relative amount of work involved in each task and the division of labor between men and women. Ask which are the most burdensome tasks. Discuss how much flexibility there is in changing the workload of particular tasks. Link the tasks and workload to tasks and activities required to be undertaken in a project; focus discussion on the participation of women.

Adaptation

This activity can be adapted to agency tasks, by creating cards depicting senior management, mid-level management, technical specialists, secretaries, lower level office staff, messengers, drivers and others.

Note: *Sample materials for conducting this activity can be found in the Participatory Development Tool Kit.*

11. Women's Lives: A Needs Assessment

Purpose

To collect information, raise awareness, and understand the priority needs of women based on their different tasks, concerns and responsibilities.

Time

1 hour

Audience

Primarily community members (women and men); also useful for trainers, project staff and field workers.

Materials needed

At least ten cards depicting women performing various daily tasks, such as:

- feeding a child
- relaxing
- visiting health center
- harvesting
- bathing
- sewing
- carrying firewood
- leading a meeting
- working in a field
- sweeping
- carrying water
- cooking
- hoeing

Blank cards should also be provided for drawing additional tasks.

How to conduct the exercise

1. Place the cards on the ground in full view of the participants.

2. Explain that the cards show women performing different tasks.

3. Ask the participants to discuss the tasks, and then categorize them into three groups: most difficult to perform; easiest; most problematic. If consensus is not achieved, note the minority opinions.

4. Allow the participants to take over the discussion as much as possible. For example, the picture of a woman and her child at the health clinic can lead to a discussion of the distance from the village to the clinic, problems in transporting pregnant women in a safe and timely manner to the clinic, and so forth.

5. Ask participants to consider which problems they can solve using resources available in the community.

Adaptation

The same activity can be repeated focusing on men's needs by redrawing cards with male figures and activities. Cards can also be drawn to show some activities being carried out by men and some by women.

Note: *Sample materials for conducting this activity can be found in the Participatory Development Tool Kit.*

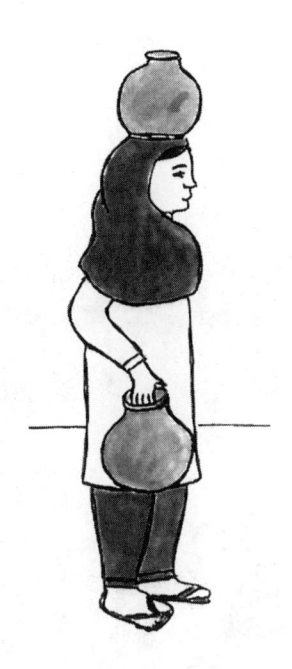

12. Women's Confidence

Purpose

To measure the degree of women's participation and self-confidence, and changes over time.

Background

Behavior displaying confidence is culturally specific and will vary across cultures. This should be kept in mind when developing locally appropriate pictures for this exercise.

This activity can be done in groups of men and women separately or together.

Time

30 - 45 minutes

Audience

Primarily community members; also useful for trainers, project staff and field workers.

Materials needed

Three cards depicting a woman in varying degrees of confidence:

- a woman too timid to enter a group meeting;
- a woman joining a water group but too shy to participate;
- a woman bold enough to talk, challenge and ask questions.

How to conduct the exercise

1. Prepare a simple pocket chart, placing one pocket underneath each of the three cards. Alternatively, pictures can be placed on the floor.[1]

---

1. A pocket chart consists of rows of horizontal pockets. Each pocket can be an envelope or made out of paper or cloth. The pocket chart is very versatile and can be used for a range of issues; to identify which water sources people use; where garbage is dumped; what are agency priorities; the most important skills of extension workers and so forth.

2. Give each participant a piece of paper, pebble or seed, and ask them to use it to vote on which woman most reflects their own feelings.

3. To determine the change in self-confidence that may have occurred over time, two sets of the same pictures can be used to depict the level of confidence before the project intervention and after.

4. Encourage discussion about what changes men perceive in women, what changes women perceive in other women, what contributed to the changes, and what have been the consequences of the changes. Before and after photographs of the situation can be used to indicate changes over time.

Note: *Sample materials for conducting this activity can be found in the Participatory Development Tool Kit.*

13. Women's Time Management

Purpose

To determine the different tasks women perform each day, the sequence in which they do them, how long each activity takes, and whether the most time-consuming activities are considered a problem.

Background

A clearer understanding of how much of a woman's day is taken up with particular tasks can help in assessing their needs. In one small town in India, women participating in this activity observed that many of their tasks were related to water and sanitation. Insights gained from this and related activities led to a dramatic increase in loan applications for latrines from village women.

Time

1 - 2 hours

Audience

Primarily community members (either all women or mixed groups); also useful for trainers, project staff and field workers.

Materials needed

Individual cards depicting different tasks which women perform in their daily routines, such as:

| cooking | sweeping | carrying water |
| harvesting | washing clothes | caring for child |

How to conduct the exercise

1. Lay the cards on the ground.

2. Ask each member of the group to sequence the cards by the order in which they normally perform the activities. Provide blank cards for drawing any extra chores the women de-

scribe. Record the similarities and differences between the responses of the participants.

3. Next, provide the group with match sticks. A full match stick represents an hour; a half match stick represents one-half hour. Ask the group to determine how long each activity takes by placing match sticks on each chore.

4. Discuss the findings with the group and let them summarize what was learned.

5. Discuss the most time-consuming chores and whether these are considered a problem. Ask the group which problems they would first like to consider and solve. Ask how they would change their daily routine if they had water closer to their home.

   Note: the exercise can be conducted with groups of men who are asked to assess and rate women's days. In addition, the pictures can be redrawn to represent a man's daily routine.

Note: *Sample materials for conducting this activity can be found in the Participatory Development Tool Kit. This activity was developed by Jake Pfohl.*

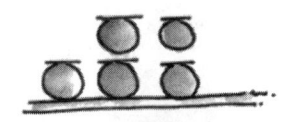

14. Women's Involvement in Operation and Management (O&M) of Water Systems

Purpose

To help women become aware of the O&M needs of handpumps or standposts in their community, and to encourage them to identify which maintenance responsibilities they are willing and able to take on themselves.

Background

Village women may not easily come forward to take on O&M responsibilities unless they clearly understand the nature of the maintenance problem and are aware of the range of tasks to be done, their level of difficulties, time needed, and so forth.

Time

1 - 1 1/2 hours

Audience

Women community members.

Materials needed

A set of 6 or more pictures of women and men or both performing a series of maintenance or management tasks such as:

- collecting money for pump repair
- water committee members educating the village women about cleanliness of pump area
- cleaning the pump site
- applying grease to the pump
- repairing the pump
- digging channel to drain waste water

- preventing water stagnation by planting trees and growing vegetables.

A set of cards representing men and women (optional)

How to conduct the exercise

1. Begin by opening a discussion of who is involved in the current maintenance system (local mechanics, water committee members, and so forth). Most of the people they name will likely be men.

2. Shift the discussion around the nature of maintenance and reliability, such as frequency of breakdowns, who reports the breakdown to whom, number of days taken to repair the system, who actually does the repair, who is inconvenienced by delay in repair, what causes the delay, and what costs are involved.

3. Next, hand out the pictures of different maintenance tasks being performed by women and men. Ask the group to study each picture carefully, analyze the relative importance of the different tasks, and discuss what the performance of the tasks involves in terms of time, labor, costs or mobility.

4. Then ask the participants to sort the pictures into 3 piles: (a) those tasks for which they are willing and able to assume responsibility (b) those which they believe should be done by men, and (c) those that can be done jointly.

   From this analysis, the role of women in the maintenance process will emerge, followed by what role women are willing to play in maintenance in the future.

Source

Adapted by ACDIL from a report by C. Rajathi, IRSENS Project Danida, Tamil Nadu, at the Danida/World Bank/ACDIL workshop in Goa, India, 1992.

C. Semi-Projective Techniques

Semi-projective techniques can be particularly useful to a researcher/facilitator when trying to elicit sensitive information from a community. These activities allow community to spill out ideas as opposed to the researcher prompting responses. Open-ended activities are fun for the community and offer a good learning vehicle for the researcher.

The activities contained in this section are:

15. Incomplete Sentences and Open-Ended Stories

16. Creating Open-Ended Stories

17. Role Playing Open-Ended Stories

15. Incomplete Sentences and Open-Ended Stories

Purpose

To enable the facilitator to elicit attitudes through the use of incomplete sentences.

Background

Incomplete sentences and stories are an interesting and very effective way to get information quickly on a variety of issues that are sensitive. Such techniques have been used to address a variety of development issues, including attitudes toward women, work, leisure, and ethnic issues.

Time

1 hour

Audience

Community members.

Materials needed

Paper and writing implements.

How to conduct the exercise

1. Read an incomplete sentence and ask one of the participants to complete it by saying the first thing that comes to mind. Participants can either be asked to respond verbally or to write down their thoughts. It is important to create the right atmosphere, make the person comfortable and avoid generating the feeling that a test is being administered. Sentences that focus on the topic of interest should be mixed up with sentences on neutral or unrelated topics.

    For example, if the objective is to elicit information on attitudes toward women, incomplete sentences could include:

A good woman should....
Men like women who....
All women are....
Women who work....
When women work together they....

Such sentences should be mixed up with neutral sentences such as:

The most important people....
In every village there should be....
Children in this community....
People who are educated are....
Children who go to school....

Another use of this technique is the presentation of incomplete stories to be completed by individuals or groups. In a participatory process, community groups can be encouraged to create their own stories.

16. Creating Open-Ended Stories

Purpose

To help extension staff understand the purpose and structure of open-ended stories, so that they can create open-ended stories on their own for use at the community level.

Background

Open-ended stories can be used in a variety of ways for understanding attitudes and practices, and for social evaluation with participants of all ages, including children. As a projective technique, open-ended stories are commonly used in psychology to evaluate cognitive capacities, moral reasoning and personal values and biases. They are also used for cross-cultural research to study cultural patterns.

Open-ended stories can be adapted for use in many different forms — as stories that are told or read aloud, role plays, drama, street theater, or puppet shows.

Time

1/2 - 1 hour

Audience

Extension staff.

Materials needed

A sample open-ended story.

Flexiflans[1], story illustrations, or other props to help the group follow the story and focus on the main characters.

A handout on the main features of an open-ended story.

---
1. Flexiflans are figures that can be put on a flannel board to tell a story. The technique was developed by Lyra Srinivasan.

How to conduct the exercise

1. Tell the group that you will share an open-ended story with them in a simple way, just as they might share it at the community level. As they listen to the story they should reflect on why it is called an "open-ended" story.

2. Read the story.

3. Leading from this discussion, engage the group in examining the format of the story in greater detail. For example, how many characters are there? Who is most concerned about the problem? How many people give advice as to the solution? How does the main character react to their suggestions? Over what period of time does the crisis occur? Draw attention to how advice is given by different people. Is there a bias in favor of any one viewpoint as presented in the story?

4. Have participants summarize the points covered in the discussion which characterize an "open-ended" story. After they have listed as many points as they can think of, distribute the handout so that they can compare it with their own analysis.

5. If time permits, invite participants to create their own stories using the guidelines in the handout.

Handout on open-ended story format

In creating open-ended stories the following should be kept in mind:

- The story should be based on local experience.

- It should pose an unresolved problem or dilemma with which people can identify.

- It should not have an obvious theme, for example selling the benefits of building a latrine.

- It should have one central character who is affected by the problem.

- The story should be relatively simple. The incident should cover a short period of time.

- It should be interesting, with some humor if possible.

- The story should include conflicting viewpoints each presented with dignity and in a realistic way, for example the traditional healer portrayed negatively, and the health clinic positively.

- The answers or solutions should not be obvious from the story. There should be room for controversy leading to consensus-building through discussion.

- Dialogue should be lively but not too complicated. The audience should be able to understand clearly what position is taken by each of the persons giving advice.

- The main character should remain unconvinced, even confused, by the advice given.

Developing effective stories quickly

- Ask children to tell you traditional stories, proverbs, and sayings. Tape record these stories.

- Ask men and women for stories.

- Ask project extension workers to write to tell you about traditional beliefs or describe a typical situation in a community that could be developed into an open-ended story.

- Ask local community people to develop stories related to particular themes.

- Try out the story with a group and change the story and the way it is used to suit evaluation purposes.

17. Role Playing Open-Ended Stories

Purpose

To help facilitators analyze the effectiveness of the role play technique for assessment of people's positions on controversial issues.

Background

Role playing an open-ended story can be a useful way to stimulate discussion and to involve the group in analytic thinking and assessment.

Time

1/2 - 1 hour

Audience

Facilitators and researchers.

Materials needed

Tape recorder and a recorded open-ended story focusing on a problem or issue that needs be resolved. (Be sure that the sound track is clear and audible).

How to conduct the exercise

1. Ask for five or six volunteers. One will play the role of a facilitator, the others will be village residents listening and reacting to an open-ended problem drama.

2. Ask the "facilitator" to introduce a recorded open-ended story with a problem to the "villagers" and, at its conclusion, guide the discussion in as nondirective a manner as possible. Remind the facilitator to try to establish an atmosphere of freedom so that the villagers will express their opinions and suggest alternative solutions to the problem posed in the recording.

3. Ask the remainder of the group to be observers. Give them specific suggestions on what to look for, particularly in the behavior of the "facilitator." For example: What kind of role

does the "facilitator" take? What method does he or she use to get the discussion going? Were the "villagers" dependent on the "facilitator" to keep the discussion moving? How were they encouraged to share their ideas, reactions, possible solutions? Did any "villagers" dominate the discussion?

4. Ask the "facilitator" to step down at this point and describe any immediate feelings about playing the facilitator role. Have the "villagers" also tell their reactions to the experience they have had. Ask the observers to give their views about what they saw happening (not what they thought or felt).

5. Lead a discussion of the use of a problem story, including other ways it might be used.

6. Encourage discussion about what is required for an effective role-play. Stress the importance of briefing people for their roles, giving them time to prepare, and allowing them to say how they felt during the role-play before inviting the observers to discuss what they saw. Ask them also what they think about the importance of disciplined observation that focuses on the specific points that the observer and trainer have chosen.

Adaptation

The problem story can be acted out by community; it can also be incorporated into street theater.

D. Games and Simulations

Some projects use games and simulations for training community and health workers. In playing the games, people reveal their thought processes, values, beliefs, and knowledge about relevant concepts and projects. The examples given here can be adapted for local use and should be program-specific.

Games and simulation activities aim to conduct data collection in an enjoyable, interactive way. Pictures, photographs, objects, role plays, puppet shows, card games, and board games can be combined in various ways to be used as research tools.

When developing simulation games, it is important to attune they fit the purpose and environment where they are played. Because games and role plays must be based directly on community realities, these games and the ways they are played will have to be developed and adjusted accordingly.

By involving community people in games, the following types of questions can be answered:

- What are the priority problems identified when communities critically analyze their current situation (self-diagnosis)?

- How do community groups perceive development potential and problems related to water supply, sanitation and health?

- Can community/project/government groups solve problems cooperatively through discussion?

- What are the implications of community perspectives for objectives and design of specific interventions (such as, in health, water, sanitation)?

The activities contained in this section are:

18. Health-Card Sorting Game

19. Health Education Game

20. Bintang Anda Game

21. Concepts of Health and Sickness: A Health Game

22. Open-Ended Snakes and Ladders

23. Health Seeking Behavior

18. Health-Card Sorting Game

Purpose

To enable agency staff to understand people's concepts of health and illness prior to developing a health program.

Background

The information learned from this activity, if desired, can be used to develop a statistical profile of the local health situation. In Indonesia, for example, data were subject to content analyses, categories were established, and responses coded. This data were then fed into a computer and a series of cross-tabulations were done to study village and gender differences. At the village level, the activity was conducted in groups and individually; it led to changes in the activities of village health volunteers.

Time

1 - 1 1/2 hours

Audience

Community members, village health workers.

Materials needed

A set of 24 cards of which 12 are pictures of practices and situations that could be associated with good health (such as use of toilet, vegetable garden, garbage collection, baby weighing, eating nutritious foods, and washing hands). The other 12 cards consist of pictures that could be associated with illness or poor health (for example, flies, garbage, defecating outside, and eating only starchy food).

How to conduct the exercise

1. Distribute all 24 cards to the participants and ask them to (1) identify what is in the picture, (2) sort the cards into two categories (good health vs. sickness) and (3) explain why they sorted them in particular ways.

2. Facilitate discussion based on these explanations. Participants, for instance, may want to resort the cards and discuss whether a practice is prevalent in the village and what problems are associated with it.

Adaptation

The activity can be used in training of staff to become more self-aware and highlight the fact that even among trained staff, there are different perceptions and disagreements.

Cards can be added that are ambiguous; for example, bottle feeding, uncovered food without flies, children playing, and so forth.

The activity can be in groups, and as a self-evaluation activity; competition can be introduced.

19. Health Education Game

Purpose

To establish the community's information base about specific health-related practices such as immunization, breast feeding, ORT and nutrition.

Background

This game can be played either individually or in groups.

Time

1 hour

Audience

Community members.

Materials needed

Two sets of cards — orange and green. The orange cards contain questions and the green cards have answers. For example, why is it important to breastfeed babies? Answer: mother's milk is best for babies.

Each card also has a picture drawn on it, so that participants who are not literate can still engage in card sorting.

How to conduct the exercise

First method: Have participants sit in a circle. Shuffle the orange question cards and set them in the center. Next shuffle the green answer cards and deal five cards to each participant. Each participant in turn picks a question from the orange pack in the center of the circle and tries to match it with the "correct" answer from a green card in hand.

Second method: Deal the cards to two groups, rather than to individuals. Follow the same procedure as above. Introduction of group interaction can make the environment more fun, with lively discussions.

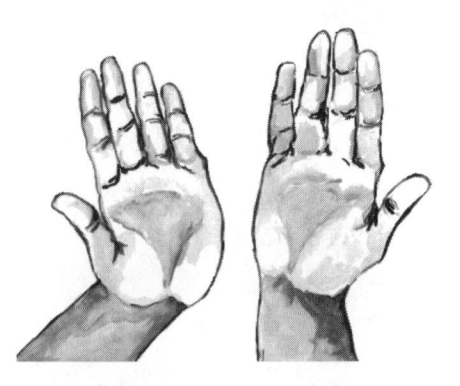

Analysis of data generated through this game can be done by the facilitator using a simple coded sheet indicating the number of correct answers given by the participants. However, comments made by the group also need to be noted to understand the reasoning underlying answers. For example, in Timor, the issue of whether breast or bottle is best for an infant always led to excited discussion. Although the 'correct' answer was supposed to be breast feeding, women sometimes stated that bottle milk was better because the composition was always the same, irrespective of the diet or health of the mother. Women also stated that if the mother did not have sufficient milk, or worked away from the home, bottle feeding was better. Once understood, village women took turns facilitating the activity with other neighborhood groups.

Adaptation

The activity can be adapted to other planning or evaluation issues, for example, financial and management implications of technology options.

20. Bintang Anda Game

Introduction

Bintang Anda — "your star" in Bahasa, Indonesia — is a family of games having the same basic structure but covering a variety of content areas. The games are very simple to play, the main focus of the activity being the discussion generated by questions, situations and role plays.

Bintang Anda presents is a game developed for use with coordination of healthy water supply. The audience for the game is agency officials and its focus is on training, planning, skill building and problem solving.

The game is led by a trained facilitator. To a great extent, the success of the game depends on three factors: the skills of the facilitator at generating good discussion; the content-related knowledge brought to the game by players, facilitator and resource persons; and the experience of the participants in group discussion and problem solving.

According to village players, some of the most useful aspects of the game involve the development of their ability to discuss and analyze problems and situations. Through the game, villagers become proud of their knowledge and understanding and less hesitant to confront persons of higher rank and status in pursuing their needs. In one case, for example, village players even invited the Chief District Officer to join in their discussions.

Villagers have also shown a great deal of creativity in developing the game in ways not foreseen by the game's originators. In one village, for instance, discussions were tape recorded. These were then played to inform and entertain people waiting to see doctors at the sub-district clinic. Doctors and nurses immediately noted an increase in questions concerning health and nutrition coming from their patients.

Source

Bintang Anda was developed originally for community development by the PENMAS, the Directorate of Community Education of the Ministry of Education and Culture in Indonesia. The information presented here has been derived from Technical Note 18 of the Centre for International Education, University of Massachusetts.

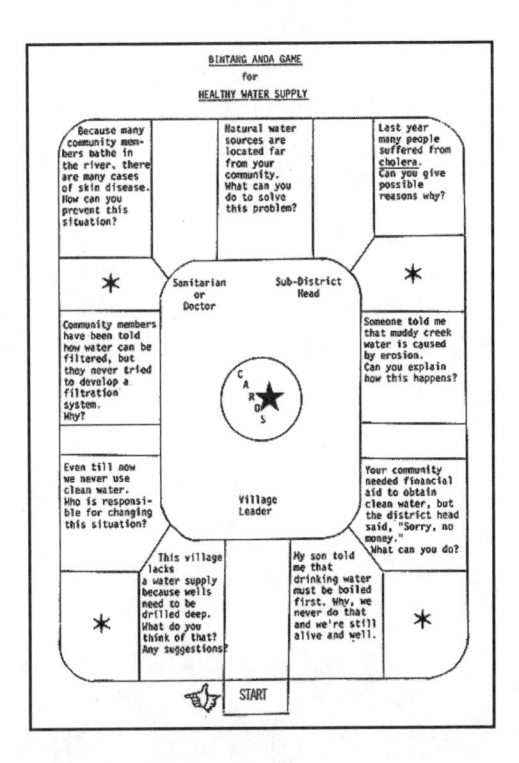

Purpose

To empower sub-district level officials to improve their abilities in:

- Understanding the situation and conditions of village communities they serve;

- Effectively utilizing existing resources;

- Managing collective agency resources for efficient problem intervention;

- Developing integrated, cooperative programs with other agencies;

- Evaluating program purchasing.

Background

The game can be played in mixed gender groups. For evaluation purposes, and to tap into gender specific knowledge and perspectives, it is useful to play the game in sex segregated groups. There should be 10 -15 participants of whom 5 - 6 are players and 3 are role players. Others present, including resource persons, serve as observers or referees who can be included in game discussion at any time.

Time

1 1/2 - 2 hours

Audience

Sub-district level officials.

Materials needed

Playing board (approximately 60 cm x 80 cm), message cards, die or number cards to determine turns.

How to conduct the exercise

1. Determine the players and their pieces.

2. Assign persons to role play village head, sanitation engineer, doctor, and sub-district head. Players take turns rolling the dice and advance around the board according to the number of steps indicated.

3. When a player's piece stops at a box containing either a message or an event description, the player reads aloud the text given in the box. The player starts discussion by offering his or her own comments and invites the opinions of role players, other players and the audience.

4. If a player's piece lands on a Bintang Anda box, the player draws one of the cards and reads what is on the back of the card. The player initiates discussion and gets the opinion of role players and observers. If another player lands on the same box, he draws a different card for discussion. (Literate members of the playing group may do the reading for illiterate participants).

5. The game continues until either one player completes the board or consensus is reached that the game is over.

6. If time permits, important points that came up during the game are pursued through further discussions.

Possible messages for Bintang Anda cards:

1. Demonstrations of water filtration have been undertaken, but villagers have not established water filtration systems of their own. What should be done next?

2. Various agencies have jointly resolved to work on the water supply problem, but no plan of action has been established. What steps should be taken?

3. We have approached informal community leaders, but they have shown little interest. Why?

4. Have our methods of approaching community leaders and members been effective? What better methods might we use?

5. What efforts have been undertaken to solve the water supply problem? What have the results been?

6. There have been efforts to inform the communities about healthy water habits, but results have been minimal. Why?

7. What are the main health problems caused by poor water supply?

8. What agencies are currently involved in working on the water supply problem?

9. Reforestation programs have been tried, but with poor results. How might these programs be improved?

10. Home gardens and household tree planting programs have been established, but many of the plants and trees have died. Why?

11. The latrine program has not taken hold in the villages. What barriers exist?

12. Why do villagers in this area still bathe, draw water, and dispose of waste in the same stream? How can we convince them to change this pattern of water use?

13. We pass instructions on to village leaders so that they can inform the community, but we have noted that their own water use habits remain unchanged. How can we get them to set a better example?

14. Regulations for water utilization have been established but they are not followed by the community. What can be done?

15. How can community participation in water supply programs be improved?

16. How can the health of the community be improved through better water supply? What resources are there for bringing about needed improvements?

21. Concepts of Health and Sickness: A Health Game

Purpose

To enable the researcher to understand people's concept of health and physical well-being (specifically sickness in infants and young children).

To understand which practices are considered health promoting and which are associated with sickness.

To evaluate to what extent knowledge about health gets translated into practice and the constraints to adopting sound health practices.

Background

This game can be played individually or in small groups depending on time and the desire for detailed statistics. If time is limited, it should be played with women from very different backgrounds. The groups can be mixed sex or segregated to compare differences between men and women.

Time

1 - 1 1/2 hours

Audience

Community members.

Materials needed

Twenty drawings of practices related to good health and illness. Four larger picture, two of which depict healthy children (one being an infant); two other pictures should depict sick children (again, one an infant).

Pictures can be cut from posters, magazines, or drawn by a local artist. Examples of practices the pictures can depict include:

- Child washing hands before eating food
- Child bouncing ball before eating

- Pictures of nutritious diet
- Picture of only starchy food
- Garbage in yard
- Vegetable garden in front yard
- Use of toilet
- Defecation outside
- Consultation with traditional doctor
- Going to a health clinic
- Breastfeeding
- Bottlefeeding
- Immunization
- Refusing immunization
- Boiling water
- Water collection from unprotected source
- Water collection from protected source
- Uncovered food
- Protected food
- Picture of flies
- Mosquitoes and stagnant water
- Parents deciding what to do with sick child who has diarrhoea

How to conduct the exercise

1. Place the four large cards of healthy and sick children on the floor and ask the participants to describe what they see. If anyone experiences difficulty identifying the pictures correctly, they should be helped.

2. The remaining cards of health practices are then shuffled and given to the participants who are asked to sort them out and arrange them below the pictures of the healthy and sick children. If a picture is perceived to be irrelevant, it can be left out.

3. Ask them to group the drawings of healthy children on one side and sick children on the other.

4. After all the pictures have been sorted out, the participants should explain why they have categorized the cards as "healthy" and "sick." If participants change their minds while giving the explanation, they should transfer the card to the correct pile.

5. Facilitate discussion through the use of follow-up questions. These can include:

    - What is happening in this picture?

    - What is he or she saying?

    - What do you think will happen afterwards?

    - Is this practice common? Is it good or poor practice? Why? Why is the practice so common or not so common?

    - How is it done in your household or family?

The facilitator should take notes or record the session.

Note: Care should be taken to stimulate frank discussion about the extent to which the practices are common in the community, the advantages, disadvantages, and problems associated with the practice. If the pictures are misunderstood or misperceived, the misperception should be corrected and people allowed to recategorize the pictures. Usually the group takes over the discussion after the initial period often working in teams monitoring each others responses.

22. Open-Ended Snakes and Ladders

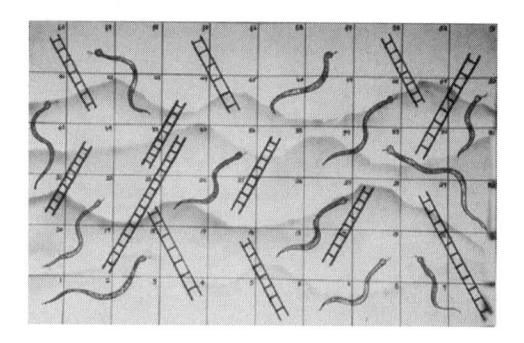

Purpose

To assist communities and health workers in analyzing the level of local health information and whether health education programs are effective.

Background

Snakes and ladders can be played either by teams or individually. Players who land on snakes go to the bottom of the snake; players who land on ladders rise to the top of the ladder. The object of the game is to reach the top of the chart first. The game generates excitement as well as learning as teams coach their dice roller on which card to pick.

This game can be conducted with children and adults; rules can be varied and made more or less complex. Participants can be encouraged to make their own rules.

Time

1 hour

Audience

Primarily community members; also useful for trainers, project staff and field workers.

Materials needed

Large chart of snakes and ladders with no directions on it.

Dice; pebbles or other markers for game pieces.

At least twenty culturally appropriate cards showing healthy and unhealthy hygiene practices. Cards can include washing hands, visiting health clinic, using a latrine, using a well, flies on food, dumping garbage, a fenced in garden and sweeping. Each card must fit within the size of the squares on the snakes and ladders chart.

How to conduct the exercise

1. Place the chart of snakes and ladders on a table or on the ground, visible and accessible to all participants.

2. Place all the cards outside the board and divide the participants into two groups.

3. Just as in the regular snakes and ladders game, this game is played with the roll of a dice. The difference comes when a team lands on a square.

4. As the dice is rolled move the appropriate spaces along the board. When a team lands on the head of a snake, the team has to correctly identify a card depicting an unhealthy practice to avoid going down the snake.

    If a team lands at the bottom of a ladder, the team has to identify a healthy practice before it can climb the ladder.

5. Because the game involves an active choice or selection of a card, it is no longer merely a game of chance or receiving health messages with time. People arrange cards and add new cards as appropriate.

Note: *Materials for conducting this activity can be found in the Participatory Development Tool Kit.*

23. Health Seeking Behavior

Purpose

To enable the researcher to understand when people take different members of the family to health facilities.

To understand gender and age differentials in use of health facilities.

To understand the weight people place on different attributes of health facilities.

Background

This activity is very versatile and in addition to discerning patterns, yields information that can be computerized. In addition to its usefulness in conducting research, it is also helpful as a lead-in to community planning and developing strategies for improving access to community-based health facilities. This activity was developed by the author for use in a recently completed World Bank Participatory Poverty Assessment

Time

1-1 1/2 hours

Audience

Community members.

Materials needed

One large picture — a simple line drawing of a health center; six smaller pictures of relevant attributes of health centers (such as people waiting, no doctor, no drugs, payment of money, distance, payment of fees, rude nurses); five mid-size pictures of members of a family: a father, mother, girl, boy, and baby. Stones or beans for indicating relative importance.

How to conduct the exercise

The activity can be conducted with individuals, or in small groups.

1. Place the picture of the health clinic on the ground. As people recognize the picture they spontaneously start making comments about their own health center. Encourage discussion and write down what people say about the health center (such basic facts, problems and good points).

2. On the side, in a column, place the pictures of the family members one below the other and say something like, "Obviously, health centers are not used for every little thing; when are different family members taken to the health center?" Focus the discussion on the kinds of illnesses and the severity of the problem which result in a family member being taken to the center.

3. Now display the pictures of the different attributes of the health center and say something like "We have already discussed the problems and good things about the center that you use. If you were going to build a health facility, what qualities would it have to ensure that you use it?" Ask the participants to distribute the ten stones to indicate their preference? Allow enough time for reflection and discussion. Note down the distribution of stones; these can later be converted to percentages if desired. The procedure can be repeated by allowing people to use only three stones to indicate their preferences.

4. Now go back to the pictures of the family members and say something like, "If resources were limited, and all the members of the family were sick, who would you try and restore to health first?" Give people ten stones and then reduce it to three. The process gets everyone involved and thinking and clear patterns emerge.

5. Ask people to comment on the patterns they see emerging and the implication of these patterns for planning health facilities and the care of different members of the family.

Note: *Pictures of attributes of the health center should reflect local realities. In Kenya, for example, a picture of an angry nurse was added to the set developed by a local artist because it was identified as an issue by local people.*

E. Technology-Related Activities

These activities are useful for gaining information on technology preferences, capacity building and letting the community know what technology options are available.

The following activities are found in this section:

24. Local Criteria for Choosing Among Technology Options

25. Pump Repair Issues

26. Understanding Pipe Water Systems Through a Model

24. Local Criteria for Choosing Among Technology Options

Purpose

To help community people clearly visualize the costs and benefits of different technology options available to them.

To assist agency staff to understand the criteria underlying community preferences for the technology.

Background

Matrix ranking is a common tool used by PRA (Participatory Rapid Appraisal) practitioners. In essence it consists of a ranking process of objects or situations on a set of criteria. This is a versatile tool which can be applied to any activity in which it is important to uncover the criteria for making choices and the relative importance of different criteria.

Time

1-1 1/2 hours

Audience

Village community members and agency staff.

Materials needed

None. However, if the discussion is to include new technologies not equally familiar to everyone, it is useful to have small models of the technologies; pictures are the next best.

How to conduct the exercise

1. When conducted at the community level, first focus on the water technologies known to local people. Later in the discussion other technologies technically feasible for the area can be introduced by the facilitator and rated by the group on the same criteria.

2. In a small group invite people to first list all the water technologies known to them. Once the list is exhausted, ask people what criteria they use to judge the usefulness of the technology. This elicits responses, like freely available, cheap technology, no technology, brings water right to the home, too heavy for women, frequent breakdown, expensive, does not affect relations with neighbor etc..

3. Depending on the size of the group, invite one or two people to draw a matrix on the ground, a large box and on the vertical axis list all the important criterion to rate a technology. People are quick at drawing symbols for the technology.

4. Next ask people to reflect on the various technologies and list them on the horizontal axis.

5. If ten criteria or technology attributes have been identified, give people, ten stones for each technology and have them distribute the stones to indicate their priority criteria influencing technology choice.

6. Once the above process is completed, focus the discussion on the attributes and strengths of different technologies, and discuss the patterns that have emerged.

7. Give people three stones to indicate their overall technology choice. People can use all three stones to give one technology increased priority. Discuss the technology preferences emerging.

8. Depending on how long the discussion has continued, you can introduce new technologies or if the discussion has already been long and involved, introduce the new technologies at a follow up meeting.

9. Gender differences can be brought out by conducting the activity in separate groups of men and women. Income differences can be used by using different types of beans, stones or sticks.

Note: *The group discussion is usually very active sometimes with much disagreement between men and women. It is important for the facilitator to point out that there are no right or wrong answers, differences do not have to be immediately reconciled, but the process is important in the community selection of the technology most appropriate for its situation.*

25. Pump Repair Issues

Purpose

To stimulate understanding among community members of the roles and responsibilities involved in maintaining water and sanitation facilities.

To initiate discussion about effective strategies for dealing with pump breakdowns.

Time

45 minutes - 1 hour

Audience

Primarily community members; also useful for trainers, project staff and field workers.

Materials needed

Three large drawings:

- A water vendor and his broken down cart
- A village scene showing a broken pump
- The same village scene with the pump repaired and functioning

Ten small cards, each depicting one factor that is important in pump maintenance and repair. For example:

- Purchasing tools
- Sweeping the pump area
- Paying the pump attendant
- People contributing money
- People speaking with an official
- Using a tool to repair the pump

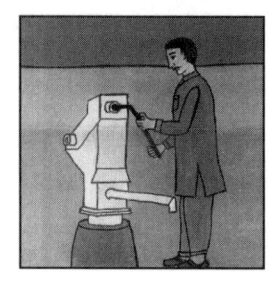

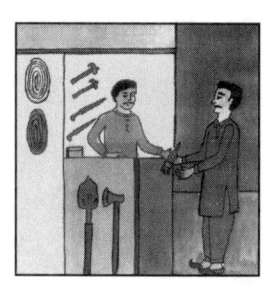

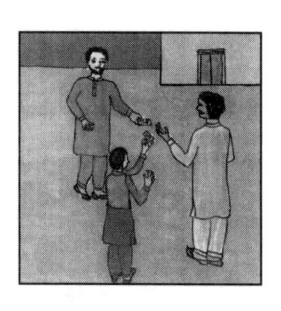

- Getting training
- Water committee meeting

How to conduct the exercise

1. Begin by holding up the picture of the water vendor with his broken down cart. Ask the participants to suggest what the vendor needs to do in order to repair the cart.

2. Next, hold up the picture of the village scene with the broken pump. Ask the participants what the community must do in order to repair the pump. Have the group contrast and compare the actions that community members must take to repair the pump with the actions the water vendor needs to take to repair his cart. Engage the participants in a discussion by asking questions such as: Who is responsible for making the repairs in each case? Who must pay for the repairs? Is a pump or a cart more difficult to maintain in good working order? Why?

3. Next, hold up the large drawing of the village with the pump repaired. Inform the participants that this drawing shows the same village one month later. Ask them to discuss the factors that could have produced the change between the two pictures. Pass around the 10 small cards to help stimulate discussion. Which steps were most important? Who was responsible for each? What action did the villagers take first? What was the order of the other actions?

4. Be sure to let the participants know that they are free to identify actions not depicted on the cards. Distribute some blank cards so that the people can draw their own key actions.

5. Ask the participants to discuss how the pump can best be maintained once it has been repaired. Who will keep it clean? Who is responsible for storing the tools? How should the pump's functioning be monitored?

Note: *Materials for conducting this activity can be found in the Participatory Development Tool Kit. The activity was developed by Lyra Srinivasan.*

26. Understanding Pipe Water System Through a Model

Purpose

To enable communities to understand, plan, monitor or evaluate the working of their pipe water system.

To assist researchers to understand the factors underlying the working of an existing system.

To uncover the decision making process, factors including land ownership and settlement patterns which influence the functioning or planning of a water system and to increase local people's understanding of the requirements of a well functioning pipe water system.

Background

Models are very effective stimuli for eliciting people's involvement. Because they are concrete and consist of at least some moveable parts, people can actually construct the ideal model or make changes in existing or planned design by actually moving pieces. This approach is useful for planning housing, water and sanitation facilities.

Time

1-2 hours

Audience

Community groups, especially marginal groups, women and the poor, engineers, system designers and evaluators.

Materials needed

Local scrap wood or cardboard can be used by local carpenters to make overhead tanks, little houses of different sizes, schools, mosques, wells, buckets, water towers, engines, valves, taps, and cardboard cut outs of people, long colored pieces of wood or pencils and straws to represent pipes of different sizes.

How to conduct this exercise

1. Place a large cloth on the ground or clear a space. Put all the pieces of the model out on the floor and tell people that different pieces represent different parts of a pipe water system. Ask people to move the pieces around and use any additional props they want from their environment to construct the existing water system.

2. Encourage discussion and participation by the group. As the group builds its system, ask the group to elaborate on the various parts of the water system, starting from the water source. This is usually a good point for people to start focusing on the history of the system, who was involved in the decision making, who is served by the system and who is not.

3. Focus the discussion on water distributional issues, how the process is managed, whether the water supply is adequate and reliable, seasonal variations, tariffs and collection issues.

Note: *The same activity can be used to plan expansion of existing systems or to plan a new water supply system. The activity is particularly powerful in getting the poor and women involved.*

F. Management Tools and Problem Identification

Theoretical endorsement of participation by management is not enough; there must be a high level of commitment to make the concept work.

Participatory activities can be useful in assessing where project management stands with regard to community participation, what they understand by it, and how they perceive it. Through these activities, management, researchers and the community can gain new appreciation of each other's perspectives and evolve a strategy to work toward the same goal.

The following activities are found in this section:

27. The Problem Tree

28. The Bottomless Pit Exercise

29. Understanding the Decisionmaking Process

30. Expectations of Community Participation

31. What is Community Management?

32. Evaluating a "Unified Vision"

    - Assessing Goals and Objectives

    - Role Perceptions

33. The "Mess" Exercise

27. The Problem Tree

Purpose

To enable community members to uncover and probe the different factors at the root cause of the problems within their community.

Background

The operating principle on which this activity is based can be applied at many different levels. In essence, it involves an analytic process of taking a problem apart and identify its root causes, as well as the spin-off effects of these causes. A Rapid Rural Appraisal (RRA) team in Pakistan describes it as "constructing problem-cause diagrams."

In the RRA model, community level problems (whether identified by farmers or observed by team members) are first prioritized and then a priority problem is written on a card and posted in the center of a board. Team members are then invited to write the causes of this problem on cards of a different color. These cards are placed below the priority problem. After identifying as many causes as possible, the effects of the problem are written on cards of yet another color and posted above the problem. In this way a complex problem-cause diagram is constructed. When finalized, team members use the diagram to draw up interventions for research and extension.

The Problem Tree exercise described in this activity is a simpler more graphic version of this model and designed to make problem analysis available at the community level.

Time

45 minutes - 1 hour

Audience

Community members.

Materials needed

Large sheet of newsprint or a blackboard; marker pens or chalk.

How to conduct the exercise

1. Start by reminding the group of one major problem identified by them in previous discussions.

2. At the top of the paper or chalkboard, note the problem, either in words or with a symbol. Next, draw a tree, incorporating the words or symbol into its trunk, branches and leaves. Show the roots of the tree reaching down in several directions. Suggest to the group that the community's problems are like a tree, and that the causes of the problems are like the roots reaching deep into the ground.

3. Ask group members to think of things that may be at the cause of the problem. As different ideas are shared, note them on the roots of the tree (either as words, or as symbols or drawings). As each cause is identified, as why it is a cause. Be sure to give participants time to reflect and discuss their responses.

4. Finally, when the full complexity of the problem and its causes comes through, ask the group to suggest possible solutions to some of the causes they have identified.

Source

Adapted from Helen Fox, *Nonformal Education Manual*, Peace Corps, Washington, DC. 1989. pp 85-86.

28. The Bottomless Pit Exercise

Purpose

To involve villagers in examining basic concepts of community management of water supply and sanitation projects and to help them apply these concepts to their own circumstances.

To assess community perceptions of their water situation and their ability or willingness to solve a range of water problems.

Time

2 hours

Audience

Village community members

Materials needed

A set of 20 to 30 blank cards; in addition, 2 or 3 cards with already-drawn pictures suggesting a problem (cattle polluting the water trough) or deficiency (lack of spare parts) related to water supply and sanitation projects.

Masking tape or thumb tacks if the cards are to be displayed on a wall or board, or pebbles to hold the cards down if the floor is to be used.

Large sheets of paper and marker pens in two colors.

How to conduct the exercise

1. Explain to the group that you wish to involve them in a management exercise that may be very useful to them in daily life.

2. Using a large sheet of paper (or the ground), draw a picture of a pit with land banks on either side. Tell them this is a pit in which we will place cards representing different problems that affect a village's water supply and environmental sanitation. Sometimes a village may have so many problems

that the pit can seem bottomless but for now it will be limited to a capacity of some 20 to 30 problem cards. Show them the sample cards with the already-illustrated problem.

3. Invite the group to brainstorm other problems to be depicted on the remaining cards. (To ensure that as many people as possible participate in generating ideas, you may wish to suggest they work in small groups of not more than 5 persons, then invite one problem/idea from each group). As each idea is offered, have it drawn on a card and place it in the pit.

4. Next ask the villagers to indicate which, if any, of these problems apply to their village. Those that do apply should be color-coded with a mark in the upper left hand corner of the card. Others will have no color code, but will remain in the pit.

5. Now label one of the two land banks of the pits "Government" and the other "Community." Invite any group member to select a card from the pit and place it on either bank, depending on whether the problem can be solved mainly by the people of the community or by the government. The volunteer who chooses the card should listen to all the options offered by the group participants. If someone feels that the solution depends on the combined effort of government and people then a curved bridge should be drawn over the pit. The card can then be placed on the bridge but it may be more to the government side or the community's side, depending on which they feel should carry major responsibility in the partnership for that particular task.

6. When most, if not all, the cards have been discussed and placed, ask the group to interpret the results. Where have they placed the major responsibility for solving local problems? Could some of the cards on the government side be taken over to the community side or at least to the bridge? If on the bridge, at which end?

7. The discussion should now be focused on the actual situation of the local community. To bring closure to the process, ask the group which one activity from the community side is, in their opinion, easiest to implement.

Source

Lyra Srinivasan, et al. 1994. "Community Participation: Strategies and Tools. A trainer's manual for the rural water supply and sanitation sector in Pakistan."

29. Understanding the Decisionmaking Process

Purpose

To encourage and stimulate a community to understand and evaluate the decisionmaking process and their own participation in it.

Time

1 - 1 1/2 hours

Audience

Primarily community members; also useful for trainers, project staff and field workers.

Materials needed

Five large cards depicting an outside official, a village official, a village or water committee, a community woman, and a community man.

Twelve smaller cards depicting key decision points within a water supply project:

- Planning
- Design
- Technology choices
- Site selection
- Fee collection
- Construction
- Maintenance

The number of large and small cards will vary depending on the local situation, and the decisionmaking process to be analyzed.

How to conduct the exercise

1. Place the large cards on the ground, and explain that each represents a person or group that may influence how project decisions are made. The exercise can be simplified by reducing the number of decisionmakers.

2. Pass out the smaller cards of project decision points, and ask participants to suggest what each card represents. Misperceptions should be clarified before proceeding.

3. Ask the participants to discuss who, in reality, determines the decision at each of these points or on each of these issues. Initiate a free-flowing discussion about the decisionmaking process, touching on key issues such as: Is there a system in place for decisionmaking and who participates in it? Who makes the decision about repairs? Who decides the size of monthly contributions? How were technology choices made? Who gets water first and who determines that? Who controls the valves that are used? Who is responsible for repairs and are they paid? How is conflict resolved?

4. When consensus is reached, have the participants place the cards next to the picture of the key decisionmaker. If there is no consensus, note the differences and continue with the process.

5. People's satisfaction or dissatisfaction with their role in decisionmaking will become clear and the discussion can then focus on what changes the community would like to see. Gender differences also become clear and can be discussed as well.

Note: *Materials for conducting this activity can be found in the Participatory Development Tool Kit.*

30. Expectations of Community Participation

Purpose

To uncover what project staff understand by and expect from "community participation."

To relate these expectations of project personnel to organizational, educational, and informational strategies utilized by the project.

Background

Everyone has a different understanding of terms like "community participation" and "community management." When expectations are unclear and different among staff at varying levels within an organization, it is difficult to change program features to enable community participation.

Time

45 minutes - 1 hour

Audience

Project staff.

Materials needed

Paper, pencils or pens, colored pencils, crayons or magic markers.

How to conduct the exercise

There are several ways to uncover people's understanding of community participation, community involvement or community management.

1. The first is to simply ask participants to write down what community participation means to them, and what community participation has meant within the context of the project.

2. Another approach is to use the word association technique. Give everyone a piece of paper and without much explana-

tion tell the participants to "free associate" and quickly write down just a few words (not sentences) that come to their mind the moment you say a particular word or phrase. Then when everyone is ready say "Community Participation." Allow a minute at most to allow them to write. When everyone is done, ask the participants to read their lists. Follow up by consolidating the lists and categorizing the lists in ways that make sense to the group. For example, behaviors of the community, strategies for project management, or positive and negative aspects. Then ask for clarification from individuals, such as: What did you have in mind when you wrote "poor people" (or self-reliance, injustice, unfair)?

3. A third approach is to ask people to draw their concept of community participation. Tell them that the quality of art work does not matter. People should first work individually, so that their unique personal perspectives are not lost. Then, in small groups, people should attempt to put their individual drawings together into a mural or a collage and explain their group concept of community participation to the other participants. When the individual drawings are being put together people can add symbols or aspects they feel are missing.

Note: *This activity involving writing statements or drawing pictures of community participation can also be used with junior project staff and senior government officials.*

31. What is Community Management?

Purpose

To facilitate management level staff in sharing ideas and understanding characteristic features of rural communities with particular reference to water, sanitation and health conditions.

To identify and examine the sets of influences or "forces" that contribute to or inhibit the process of promoting community managed water and sanitation facilities.

To select priority issues for action.

Background

Senior personnel no less than line staff need to arrive at a common understanding of what "community management" is and why it is important. Reaching this understanding through a participatory activity has the added benefit of providing first hand experience of the richness of ideas generated through the participatory process.

Time

1 - 1 1/2 hours

Audience

Management level staff.

Materials needed

Large sheets of paper, markers pens, scissors, glue and tacks.

How to conduct the exercise

1. Invite the participants to form two groups. Ask that each group identify the features of a typical rural community in their area. Emphasis should be given to water, sanitation and health conditions, but critical socio-economic and cultural factors that serve either as propellants of positive change or as hindering forces should be recorded, as well.

2. Once this activity has been completed, each group should prepare a dramatic representation of the process of achieving optimal community management. Participants should assume roles of key figures that influence the achievement of the community management goal. Similarities and differences in the presentations should be critiqued by the whole body of participants following the presentations.

3. Ask the groups to then apply the "forcefield analysis" technique to brainstorm and identify clearly the factors that contribute to or hinder the empowerment of local communities in decisionmaking and management. The lists of factors should then be consolidated by the participants.

4. This analysis should lead to consideration of components of an action plan to achieve the community management goal.

Source

Adapted from a workshop conducted by Ron Sawyer, World Bank, Kenya.

Note: *Materials for the Force-Field Analysis technique referred to above can be found in the Participatory Development Tool Kit.*

32. Evaluating a "Unified Vision"

Projects cannot work smoothly or achieve success in applying principles of community management when project personnel have differing views of the project's overall purpose, goals and objectives. Officials and agencies which are in close contact with communities may have different ideas about community initiatives and participation. Understanding roles, such as those of the communities and other project personnel, is also important.

It is important, therefore, to assess whether project staff have attained a unified vision of operational strategies. It is equally useful to compare staff perceptions with the perceptions of the community.

Two exercises described below can be used in assessing the degree to which a unified vision exists within the project. Both techniques can and should be used at agency and community levels. For follow-up action, it is important that the perspectives are not only discussed within each group but across groups. This may "unblock" communication problems and lead to adoption of more effective strategies.

Assessing goals and objectives

Purpose

To evaluate the extent to which the perceptions of project personnel and the community about the project's goals coincide and are mutually supportive.

Time

1 hour

Audience

Project staff and community members.

Materials needed

Paper, pencil, tape recorder (optional).

How to conduct the exercise

Depending on the trust and comfort level of the group either of two procedures can be used in this exercise. The discussion is usually lively as people are either surprised and pleased by the shared vision or surprised at the differing perspectives.

1. Introduce the activity by telling the participants that the goal of the meeting is to try to understand and record in some way why the project has been successful. Since they have been most directly involved in it, they are the best people to answer these questions. Everyone has unique perspectives and experiences with this project, and it is important not to lose these perspectives.

2. Each participant should write down answers and feelings regarding one or two of the following questions:

    - What does "success" mean? How is success measured within the project?

    - Do you think the project is successful? Why? or Why not? Please explain.

    - What are the most important factors that have contributed toward its success or lack of success? What factors have contributed to that result?

    - What are the overriding goals of the project? What is the project trying to achieve or demonstrate?

3. After the writing is concluded, a discussion should be held.

4. Alternatively, participants can be divided into small groups and asked to think ahead about what their agency or project may be able to achieve at the end of a given span of time. Using their imagination, they should project a success-image of this project or institution and compare notes with other groups.

5. If a project document exists, it is useful to introduce the goals and objectives as in the project document and discuss them. What do people think about these objectives? Are they clear, valid? Have the objectives become broader?

Adaptation

The exercise can be adapted to the village level through small group discussions or individual interviews. Writing should be kept to a minimum.

Role perceptions

Purpose

To understand and clarify perceptions about people's roles in different sectors and at different levels.

Background

This exercise helps to overcome misconceptions or unrealistic expectations that staff within a ministry or agency may have of one another. It is particularly appropriate when training a group that is multi-level within one ministry or organization, and thus may know very little about each other's day to day experience. This exercise was developed to help participants "see themselves as others see them" and to create genuine willingness to compare and discuss mutual role expectations.

Time

1 hour

Audience

Government and agency staff

Materials needed

Newsprint

How to conduct the exercise

1. Divide the group into subgroups according to their professional speciality or level.

2. Ask each subgroup to define their own role and the roles of one group immediately below and one group immediately above their own. Members should write out the roles as they perceive them.

3. Have all the groups post the results in horizontal rows, one under the other, in such a way that the roles of any one category as seen from different perspectives can be compared in a vertical direction. Participants can walk around and observe the perceptions of all the other groups.

4. Discuss in the large group the discrepancies in views about each role and the implications for future team planning. Ask them for suggestions of how to work together more effectively now that they understand each other's roles better.

Source

Lyra Srinivasan. 1990. *Tools for Community Participation.*

33. The "Mess" Exercise

Purpose

To assist personnel of community-based projects to review and assess the obstacles to community participation observed in their work experience.

To enable them to categorize and prioritize these constraints so that a clear picture can be obtained of the program components which need major attention (with a view to restructuring field strategy and planning future courses of action).

Background

Field supervisors and extension staff often fail to review and report their project experience analytically for subjective or practical reasons. They may be overburdened with reporting requirements on primarily administrative matters; or they may consider their experience to be of marginal interest to others, especially to "higher-ups." Yet their insights are valuable for correcting the course of managerial and operational strategies. A simple group activity, such as the following "mess" exercise, can show participants the importance of their inputs to the planning process. It also gets them thinking about the possibility of similarly involving community members in investigation and assessment of their daily life situation.

When applied at the Danida/ACDIL Workshop, this activity clearly brought out several common denominators and important differences in the experience of participants from different states of India. Despite the differences, the largest number of constraints identified through the activity turned out to be under the category of administration. On the basis of this exercise, participants formed recommendations for improving policies for community based programs.

Time

1 - 2 hours

Audience

Project staff.

Materials needed

Blank squares of paper, approximately 5 cm x 5 cm in size and at least 5 squares per participant plus a few extra.

Blackboard or flip chart paper.

Marker pens, other writing material, adhesive tape.

How to conduct the exercise

1. Explain the purpose of the activity.

2. Distribute a few sheets of paper to group members; position extra sheets within their reach.

3. Ask the participants to write down different constraints they have met while applying participatory approaches. Each constraint should be written on a separate piece of paper. Through this process, a large quantity of assorted, undifferentiated data on constraints will be produced.

4. The next step is to sort out the "mess" by clustering the constraints under different categories. Ask each participant to read aloud the text on one piece of paper and place it on the board after declaring the category to which it belongs (administration, funding, community attitude, political influence, training deficiency, and so on). Discussion of whether or not the category fits should be encouraged to deepen understanding and reach consensus. New categories may need to be created during this process.

5. When all papers have been posted on the board, the group should assess which category has the largest number of constraints and which has the least.

6. The focus should then shift to ways of overcoming constraints.

Source

ACDIL (Academy for Community Development and International Living), activity introduced at the Danida/World Bank/ACDIL follow-up workshop, Goa, India, 1992.

SECTION THREE

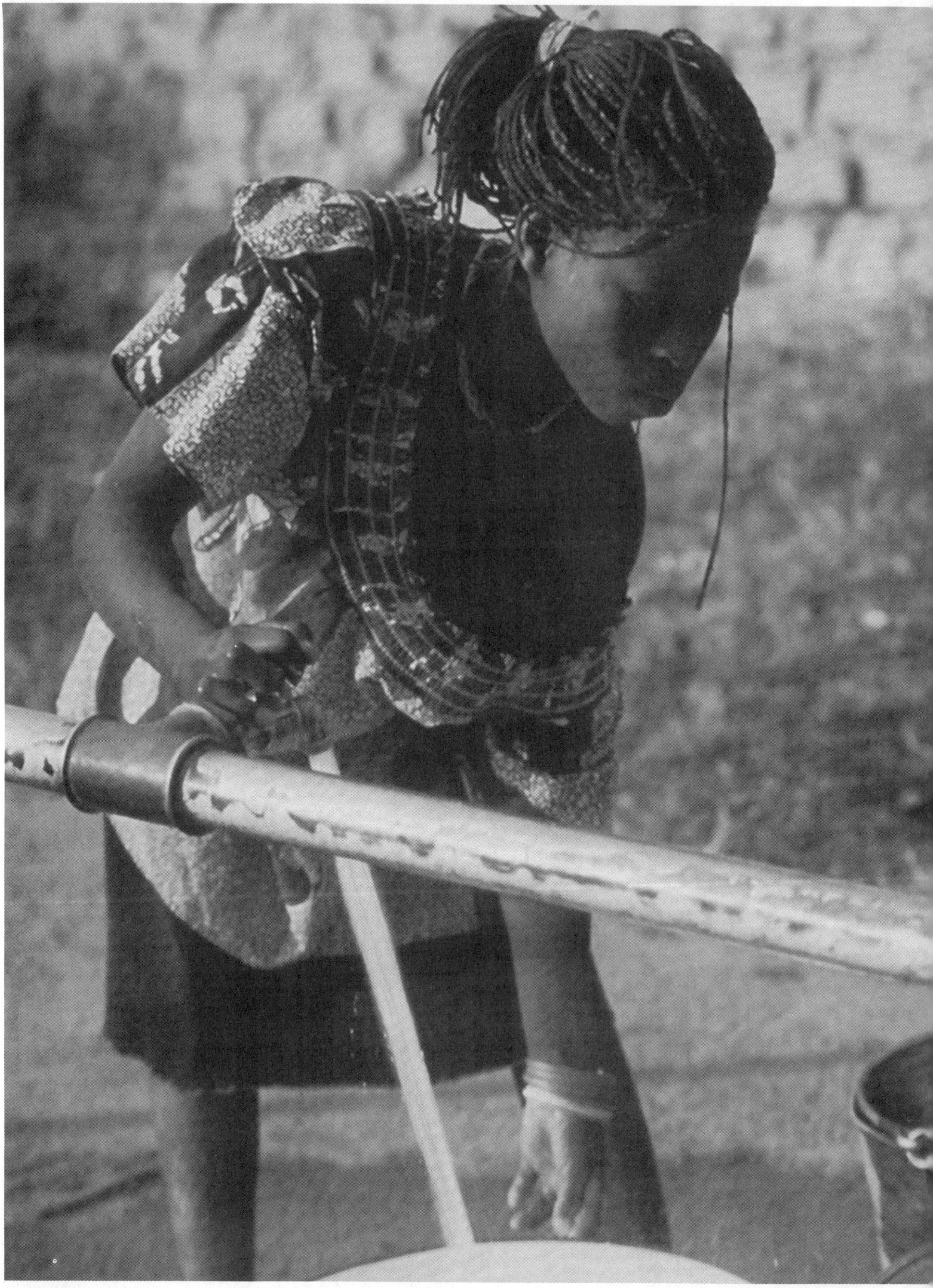

Chapter 11

Participatory Research Checklists

Checklist purposes

A checklist is a flexible tool that can serve a variety of purposes. It can, for example:

- Provide an overview of the problem to be studied — the nature of the problem, its dimensions, and the specific components on which research should focus;

- Define the scope and points of emphasis of a feasibility study (prior to project identification and screening for possible financing);

- Facilitate decisions on strategic components that need priority attention and those that do not;

- Highlight needs and goals, constraints and resources, strengths and weaknesses, which require special attention in order to ensure the sustainability of project outcomes;

- Help establish which inputs are considered by project managers, field supervisors, extension agents, and others to be most crucial to the success of the project;

- Provide the researcher with a certain amount of security in conducting focused interviews on a selected topic, problem or issue, and, at the same time, allow flexibility in adapting the interview to serve specific requirements;

- In the hands of an experienced and innovative researcher, it can open up dialogue between the field worker and community people or agency staff. Checklist items can be modified, dropped, reordered, or supplemented to suit the needs of the group with which it is used. Thus, it can serve to initiate an exchange of ideas in a fluid manner and with a high degree of spontaneity, without sacrificing depth.

Types of checklists

Checklists can be structured in different ways, such as a list of questions or of points. The levels of detail and accuracy of informa-

> **Chapter Contents**
>
> *This chapter explores the following issues:*
>
> Checklist purposes
> Types of checklists
> Alternative ways of using checklists
> Information needs for designing sustainable community-based programs
> Community level checklists
> Checklists combined with other activities
> - Community setting
> - Economic factors
> - Health and hygiene practices
> - Strengths analysis of community level organizations
> - Needs assessment
> - Existing water situation
> - Excreta disposal facilities
> Agency checklists
> - Pre-planning stage checklist
> - Management of services
> - Dimensions of project organizations that are important for sustainability analysis
> Gender analysis

tion needed will vary depending on the use of the data. If a checklist is being used to help provide an overview of the community's water supply and sanitation situation and serve as a basis for community mobilization and planning supportive institutional resources (such as technology, manpower, and equipment), the information considered pertinent will be lengthy and detailed. In the interest of avoiding information overload, the checklist should be streamlined to its essentials, keeping the intended purposes clearly in view.

There is nothing that binds the researcher to adhere to the checklist. On the contrary, it offers scope for redefining data collection priorities through discussion at different levels. Selectivity through a participatory process is one way of fine-tuning the checklist. The sample checklists included in this chapter illustrate a wide range in structuring, focus, and coverage: from those that are very broad in scope and detailed and comprehensive in coverage, to those that focus on specific issues or aspects of projects and examine them in depth.

To get an overview of the situation in which a community-based project evolves, the generic screening checklist must not overlook other factors even at the risk of throwing the net too wide. Choices can then be made for specific purposes, resulting in shorter and more focused lists of items for data collection, such as for purposes of a feasibility study, a water-use study, or a training needs assessment.

These topics should be kept in mind when reviewing the scope of items listed in the "Community Level Checklist," which covers the following topics:

- Village setting, infrastructure, demographic factors;
- Economic factors, social systems;
- Health and hygiene concepts and practices;
- Village organization;
- History of community participation;
- Needs assessment;
- Traditional roles of men, women and children;
- Available technology and resources;
- Environmental sanitation.

Within the general framework, a few examples are given in the section on "Checklists Combined with Other Methods" on how specific components from a broader checklist can be selected for more focused and analytic study.

Alternative ways of using checklists

Although checklists are commonly used as guidelines for interviews, they are also useful as a basis for planning other forms of data generation, in particular for participatory investigations at the community level. Even highly creative activities such as map-building can utilize checklists as a means to assess whether the information generated is comprehensive and accurate. The section on "Checklists Combined with Other Methods" identifies alternative ways of generating data in line with the suggested list of topics of direct concern to project planners and managers. It also points out a number of notes on what researchers should avoid when using the checklist.

Neither of the two examples of checklists given in "Community Level Checklists" or "Checklists Combined with Other Methods," however, is intended to fulfill the need for a comprehensive conceptual framework to guide the design of the project's strategy. Such a framework is found in the section on "Information Needs for Designing Suitable Community-based Programs." The focus of this section is on the dynamics of agency and community interaction, starting with indicators of community motivation, readiness and ability to shoulder partnership responsibilities such as cost sharing and management capabilities. In this conceptual framework the role of the agency, as supporting institution, is also closely examined in terms of its responsiveness to an impact on local demand.

In addressing the needs for planning sustainable programs, data collection can be organized in two stages: the first focused on defining the need for the project, and the second on choosing appropriate technology and management, both within the community and in the external resources.

Information needs for designing sustainable community-based programs

The focus of this section is on the dynamics of agency and community interaction, starting with indicators of community motivation, readiness, and ability to shoulder partnership responsibilities such as cost sharing and management capabilities. In this conceptual framework the role of the agency, as supporting institution, is also closely examined in terms of its responsiveness to an impact on local demand.

The first category of questions focuses on assessing community demand and agency capacity to be responsive. The second category focuses on technology and management capacity both within the community and in the external environment for support or co-management arrangements. The information needs in regard to each of these two stages is summarized below.

Category 1
Matching community demand with agency supply:
Criteria for selecting communities and agencies

1. Who are the users?

2. What is the willingness to pay among users (women and men)?

3. How do the following factors affect willingness to pay?

   - Household characteristics
   - Satisfaction with current facility
   - Characteristics of new facility
   - Characteristics of supplier

4. What is the ability and willingness of community groups (women, men, mixed) to commit resources and organize?

   - Functioning of community groups and networks
   - Local leadership
   - Willingness and skills of groups to manage WSS

5. Are the poor targeted?

   - What is poverty in the local context?
   - Who is poor?
   - Are the poor represented in community groups?

6. What is the capacity of the agency to assess and respond to expressed demand?

   - Mandate of the agencies, goals, objectives
   - Agency culture, rules and regulations, planning and evaluation tools
   - Legal framework
   - Rules regarding investment decisions, financial flows
   - Staff profiles, skills, personnel evaluation criteria
   - Level of administrative and fiscal decentralization
   - Community outreach, field presence

Outputs

1. Nature of the felt need

2. Importance of the need

3. Who perceives the need:

   - Wealth and power
   - Gender
   - Age
   - Spatial marginality
   - Groups and agencies

4. What is the willingness to pay for different service levels

5. Social organization of groups, membership, leadership, functioning, and skills of groups

6. Willingness of users to make other possible commitments

7. Extent of poverty and poverty characteristics

8. Capacity of proposed agencies to respond to demand, strengths, and weaknesses

9. Appropriateness of agency to support local group management capacity

10. Institutional arrangements for facilitation and delivering

Does the process of data collection contribute to local capacity and ownership?

Policy, design, and strategy outcomes

Outline the program policy, design, and strategies

- To assess demand
- To respond to user demand or create demand
- To reach the poor
- To empower women
- To strengthen community capacity and management
- To institute policies, legislation and mechanisms to enable agencies to be responsive (including fiscal decentralization)
- To enable participation of other agencies, NGOs, and the private sector

Does the process of determining outcomes create local ownership?

Category 2
Designing sustainable management systems:
Criteria for determining technology options and local management capacity

1. Technology preference: what is the service level desired by users?

   A. Is technology desire based on:
   - Experience with existing facilities
   - Exposure to other options
   - Perceived benefits and uses
   - Perceived disadvantages, problems

B. How is convenience rated based on access?

- Ownership of land, water, technology
- Availability at all times
- Waiting time/queuing

Based on distance?

Ease of use?

- Physical strength
- Posture, position needed for use
- Ease of access (entry/exit, terrain)
- Spatial spread within facility (distance of water from toilet)
- Location and orientation

C. Is multiple use of facilities important?

- Interest in multiple use
- Environmental and technical issues
- Features needed to increase multiple use

D. Is the technology affordable?

- Total capital costs
- Total operation and maintenance costs
- Subsidy needed in relation to WTP

2. What is the local management capacity for effective operation and maintenance?

A. How reliable is the technology?

- Life of technology and components
- O&M requirements to minimize breakdown

B. Who will be responsible for repairs?

- Level of skills needed
- Tools needed
- Availability of spare parts, inputs
- Costs of parts, other inputs
- Backup technical support needed
- Organization and linkages needed for repairs

C. What is the organization of community and agency and how well are they functioning?

- Communication network
- Leadership
- Membership, structure and functioning of groups and agency

- Autonomy and accountability of financial resources (incentive to do O&M, and repairs)
- Accountability to users
- Conflict resolution mechanisms

Outputs

1. What are the range of technology choices?

2. What are the desired service levels?

3. What are the cost sharing options?

4. What are the responsibility sharing options?

5. What are the design features for multiple use?

6. What are the local O&M management requirements?

7. What, if any, are the technical support services needed?

8. What are the systems for communication and outreach?

9. What are the features of responsive and accountable community groups and agencies?

Does the process of data collection contribute to local capacity and ownership?

Policy, design and strategy outcomes

What are the policy, design, and strategies:

- To ensure service level choice based on WTP and ability to organize and manage O&M of the system
- To achieve accountability and relative autonomy of decisionmaking by community and agency level
- To ensure accountability to user for quality of service provided
- To facilitate two-way information flow and communication, linkages between different levels
- To strengthen capacity and legal status of community groups
- To strengthen capacity of appropriate agencies to deliver needed support (materials and empowerment)

Does the process of determining outcomes contribute to local capacity and ownership?

Community level checklist

1. Community setting

    - Physical terrain
    - Physical distance from national, regional, provincial, district capitals, and towns
    - Degree of accessibility during the whole year by road or boat
    - Rainfall, rain season
    - Total population, total area
    - Ethnic, religions, language groupings
    - Division of community into smaller units and sub-units
    - Settlement patterns

2. Infrastructure

    - Major and minor roads within the community
    - Electricity
    - Markets and shops, access to markets
    - Cooperatives
    - Pre-primary, primary, and secondary schools
    - Health care facilities, physical accessibility
    - Presence of government, non-government extension workers
    - Other ongoing development activities, projects, agriculture forestry, livestock, irrigation, formal and non-formal education, health.

3. Demographic factors

    - Total number of households
    - Household composition
    - Special features of household composition: female-headed households, de jure, de facto, presence of non-kin members, relationships
    - Adult/child ratio per household
    - Number of children younger than 5 years per household
    - Long term and seasonal or cyclical migration patterns

4. Economic factors

    - Physical types of dwellings, their conditions, layout
    - Building materials used, types of roofs
    - Space available inside and outside house
    - Primary source of income
    - Secondary source of income
    - Seasonality of occupations
    - Income levels
    - Availability of cash, seasonality
    - Presence of consumer goods: radios, bicycles, watches, tape recorders, televisions, etc.
    - Ownership of livestock and production tools
    - Land tenure or ownership patterns

5. Social factors

Major social and political factors which divide communities are

- Caste, religion, clan, ethnic, political
- Marginal or disadvantaged groups, their characteristics, poverty, people without land or assets, female-headed, minority groups
- Concept of poverty

6. Health

Major diseases presenting mortality and morbidity patterns including

- Infant and child mortality
- Number of children born per woman
- Malnutrition in children
- Staple diet food consumption
- Intra-family distribution of food

7. Community organizations, formal and informal

- Type of organization, its objectives, legal status
- Structure, laws, by-laws
- Decisionmaking process
- History of organization
- Financial management
- Past and present strengths and weaknesses
- Community access and representation
- Past achievements
- Present activities
- Community perception of institutions and leaders

8. History of community participation

- Traditional mechanisms for community participation
- Context, form, origin of stimulus of community participation
- Purpose, strengths, weaknesses, who makes what decisions
- Role of men, women and children in community participation
- Leadership patterns

9. Needs assessment

- Perceived priority needs
- Differences by gender, social, economic status
- Perceived need for water improvements, willingness-to-pay
- Perceived need for sanitation improvements, willingness-to-pay
- Perceived need for health improvements, willingness-to-pay

10. Traditional roles of men, women, and children

    - Task and role analyses by age and sex
    - Do's and don'ts of men, women and children
    - Task analysis, gender-bound tasks
    - "Difficult to do" tasks by age and gender
    - Men and women's role in family and community level decisionmaking
    - Role of children in child care

11. Available technology and resources

    - General level of technology use
    - Presence of people with specialized skills
    - Builders, mechanics, welders, communicators
    - Availability of technology-related input, sand, wire mesh, water, building materials etc.

12. Education and exposure to media

    - Literacy levels of men, women, and children
    - Pictorial literacy
    - Exposure to mass media, radio, T.V., pamphlets, etc.
    - On-going educational activities: formal and informal for adults and children
    - Traditional communication modes and channels: stories, riddles, puppets, theaters, drama, dances, etc.

13. Existing water sources

    - Location
    - Type
    - History, beliefs, myths, rituals, and ceremonies
    - Physical and social access
    - Water quality
    - Functioning
    - Reliability, seasonality
    - Operation and maintenance
    - Use and transport of water
    - Perceived advantages and disadvantages of water sources
    - Disposal of waste water
    - Use by animals

14. Environmental sanitation

    Existing excreta disposal facilities:

    - Location
    - Type
    - History
    - Physical and social access
    - Quality
    - Functioning

- Operation and maintenance
- Use by age and sex
- Perceived advantages and disadvantages
- Disposal of infant and child excreta
- Attitudes toward handling of excreta, especially children's excreta
- Cleansing and ablution practices
- Beliefs, taboos, preferences
- Problems related to disposal of human excreta

Household cleanliness:

- Disposal of household waste
- Disposal of waste water
- Overall cleanliness of houses

15. Health and hygiene practices

Concepts of good health and illness as applicable to men, women, and children:

- Perception of diarrhea as a problem
- Treatment of diarrhea
- Food handling and storage practices
- Water handling and storage practices
- Hand-washing practices
- Cleansing materials used
- Personal cleanliness practices
- Location

Checklists Combined with Other Activities

1. Community setting

Purpose:

To obtain information about community setting (village, urban, peri-urban community, slum, or squatter area) including demographic, development level, and infrastructure available.

Checklists:

Community setting, infrastructure, demographic factors.

Methods:

Mapping (rough) by community people; agency staff or key informant interviews (2-4) (chief officials, community man, community woman; village register or census in groups as secondary sources.

Note:

Do not get carried away with the degree of detail. Resist the temptation to obtain information through house-to-house surveys. The information on adult and child rates per household can be obtained through a "mini survey" or through group surveying. In the presentation of results, key features which will affect project design and implementation should be highlighted, such as village geography, the duration of the rainy season, and how development activities are functioning.

2. Economic factors

Purpose:

To obtain information about the economic conditions and means of livelihood of households with specific reference to gender differences and households in poverty.

Checklists:

Economic factors, social factors

Methods:

Group discussions, key informant interviews, walking surveys, group census, demographic pebble, who is poor, what is wealth.

Note:

It is common to try to obtain precise quantitative information which can be reported as percentages in tables (such as number of men in various occupations, number of households owning watches or cattle, and types of households with types of rooftops). Unless there is clear justification for this type of information, all attempts at categorizing each household to develop scales of wealth should be avoided.

It is more useful to know how home improvement decisions involving costs are made within the family than simply record the number of improved houses. For example, it is more useful to note that savings by men are invested in house improvements, while savings by women are invested in informal savings networks, than to find that 23.8 of houses have glass windows.

In identifying pockets of poverty, key informant interviews should be conducted away from the community center, and closer to where poverty conditions exist. Sometimes community officials may deny, ignore, or be unaware of marginalized groups.

3. Health and hygiene practices

Purpose:

To understand prevailing concepts of good health and illness and prevalence of health promoting behavior among men, women, and children.

Background:

Typically studies try to collect primary data on diarrhoea episodes, infant mortality, and morbidity. It is extremely difficult to get useful information quickly on these subjects. Hence, it is strongly recommended that no attempts be made to obtain these statistics. Use should be made of existing data from health clinics and published sources, unless a health impact study is being conducted.

On the other hand, in order to design an effective project which includes hygiene education, it is important to understand local health related concepts, such as concepts of good health, sickness in children and in infants, and how diarrhea is perceived and treated. Other health related practices include food and water handling, beliefs about good nutrition and local diet.

Checklist:

Concepts of good health, illness, environmental sanitation

Methods:

Observation; health games with pictures, simulations, stories, role plays; or key informant interviews with young children, traditional healers, mothers, primary health care staff.

Note:

If trust is established, indigenous knowledge systems can be tapped through informal conversations, pictures, and games. However analyses of such data should be systematic so that the bizarre, different, or exotic do not get over-reported.

4. Strengths analysis of community level organizations

Purpose:

To understand the strengths of existing community level organizations and networks, both formal and informal.

Background:

Without building strong local organizations, sustainability cannot be achieved. Merely listing activities of different organizations and their formal structure and function is a beginning but not sufficient. In the past we have tended to focus too much on weaknesses, or "deficit analysis" which, though useful, is not helpful in designing strategies to achieve sustainability.

Even though "modern" organizations may have been created as the lowest level of government agencies, the functioning and management of these organizations is still based on indigenous management practices. For example, indigenous means for conflict resolution. These cultural aspects which determine how an organization is run are important to understand. This is sometimes easier to discern in informal or traditional organization, e.g. traditional forms of community cooperation; informal women's revolving credit societies; men's work groups; burial societies etc. vs. village office and treasury; formal cooperative, literacy group.

It is especially important to focus on the strengths of these groups so that they can be used as a starting point rather than focusing only on the weaknesses of the existing formal organizations.

Checklists:

Community organizations, history of community participation

Methods:

Informal conversations and interviews; discussion with members and leaders especially some traditional leaders;

Note:

If time is short, merely list the formal institutions but explore the informal management systems and leadership patterns operating in the formal and informal organizations. If the formal structures are generally known, these need not be addressed in each community setting.

5. Needs assessment

Purpose:

To understand people's perceived needs and the ranking of water and sanitation within the framework of priority needs.

Background:

Direct questioning by outsiders, such as "Is water a problem?" leads to answers considered socially desirable but not necessarily true. If communities are expected to manage systems, it is important to know if water and sanitation are really perceived to be problems so that agencies can give priority to communities where water and sanitation is a perceived priority problem.

Checklist:

Needs assessment

Methods:

Indirect methods: force field analyses; story with a gap; rating of family problems; gender specific group discussions; use of flexi-flans; drawings.

Note:

If participatory methods are being used by extension workers who are sector specific (health or sanitation only), and if priorities that emerge are other than water and sanitation, extension workers must try and ensure follow up action by the competing government departments. Follow up is essential to maintain the trust of community people.

6. Existing water situation

Purpose:

To assess the existing drinking water situation

Background:

There are many issues of importance ranging from history of water source, ownership, access convenience, use, etc. Hence, the checklist is organized by source characteristic.

Checklist:

Existing water sources

Methods:

Site visits, key informant interviews with owners, people living near sources (women and men), and mapping.

Convenience of water sources

Methods:

Observation, group discussion, key informant interviews with users and non-users

Functioning and use

Methods:

Observation, water quality tests, measure water flow in liters/minute, group discussion especially with women and people living near source.

Note:

Sometimes the same source may have different names and local people may initially mention only the major sources. In a study in Indonesia, a village which initially was reported to have three sources actually had over twenty springs.

It is important to cross check information on breakdown or facilities with community people if outsiders are gathering information.

7. Excreta disposal facilities

Purpose:

To understand the existing situation regarding excreta disposal and the factors influencing the situation.

Background:

In order to develop effective strategies to promote improved sanitation, it is crucial to understand the social and cultural factors underlying existing practices. It is extremely difficult to obtain useful information through direct questioning, hence, open ended, indirect, and participatory methods are strongly recommended.

Checklist:

Existing excreta disposal facilities, latrines, sewage systems

Methods:

Pocket-chart self survey, open-ended questions of advantages and disadvantages of various options; gender specific discussion using models with moveable parts; some projective techniques using pictures as stimuli of improved and unimproved environmental sanitation, mapping.

Note:

Although it is important to have some idea of the number of people (gender specific data) using various options, if time is short it is more important to do a few in-depth interviews or discussion groups to understand the situation rather than attempt a full-scale household survey.

Agency Checklists

Community participation in planning and decisionmaking

This checklist is the first part of a much longer instrument in which the potential role of the community in project development as a whole is examined. The issues raised are limited to the project's initial stage but serve to illustrate the type of critical questions which need to be raised.

1. Pre-planning stage checklist

In this important stage, it must first be decided how and to what degree the community can participate. Critical questions include:

- Is there a legal framework which permits community participation?
- What has been the background of community participation in the country and particularly in the region of the project?
- What is the likely level of "social readiness" for the changes envisaged and for the desired level of community support?
- What governmental and non-governmental organizations are concerned with water supply and sanitation, community participation, and the involvement of women?
- Who can assist in preliminary designs of community participation strategies?
- What is the variation in the country or region in terms of cultural traditions, languages, and felt need for improved water supply and sanitation?
- How will technological solutions influence levels of acceptance and community participation?
- What is the political climate which supports or constrains community participation?
- How can existing social or developmental structures be best used in the new project?

Source: *IRC International Water and Sanitation Center Community Participation and Women's Involvement in Water Supply and Sanitation Projects.* Occasional Paper Series 12. The Hague, Netherlands.

2. Management of services

This checklist is intended to serve as a set of guidelines on management and maintenance systems of projects. Its aim is to (1) document the components of sustainable maintenance systems, (2) highlight lessons learned, and (3) focus on elements critical to system sustainability and replicability. Because this checklist is intended to serve as a guideline, not all issues need to be addressed and additional items may be inserted.

1. Summary of project

    - Project duration
    - Location
    - Number of people served
    - Main project components
    - Type and number of systems installed

2. System evolution and description

    - Type of maintenance system: village maintenance, local artisan, centralized system
    - Type of maintenance: preventive, curative, or both
    - How and when was system established?
    - How was commitment to take responsibility for O&M achieved?
    - What is management process? who makes decisions and how?
    - How is management process established?

3. Reporting process

    - Who reports breakdowns to whom, and how?
    - Average delay between breakdown, reporting, and repairing
    - O&M records maintained, type of information collected, by who

4. Repairing and servicing

    - Who repairs and services what?
    - Who selects repairers and how?
    - What tools do repairers have and how were they procured?
    - Who trains repairers and how?
    - Number of systems served per repairer
    - Legislative framework for allocation of responsibility and quality control

5. Payment

    - Repair costs: who pays, how much, who sets rates (labor, transport, and parts)
    - How are payments made? who collects money? who accounts for money? how?

- Levels of cost recovery and subsidy (how much and how), is the system affordable to users?
- Warranties and service contracts

6. Spare parts

    - Who makes spare parts?
    - Who procures spare parts? who distributes and sells spare parts?
    - Prices, price controls and mark-ups

7. Subjective assessment

    - Percent of systems functioning and basis of assessment
    - Potential problems of management process and external back-up support provided (technical and extension); long-term monitoring process: by whom and how
    - Potential for system expansion, rehabilitation and upgrading
    - Spare part availability
    - Sustainability and replicability: key issues and constraints

Source: Jennifer Sara, UNDP-World Bank Water and Sanitation Program, 1991.

3. Dimensions of project organizations that are important for sustainability analysis

1. Input factors (quality and quantity)

   - Staff
   - Administrators
   - Facilities and equipment
   - Appropriate technology
   - Finances

2. Output factors

   - Quantity
   - Quality
   - Level of demand for outputs
   - Rate of use of outputs if they are factors of production
   - Public and private or both

3. Technology factors

   - Standardization
   - Repetitiveness
   - Control by rules
   - Scale
   - Sophistication of technology
   - Appropriateness for available staff
   - Dependency on foreign experts

4. Structural factors

   - Formalization
   - Centralization
   - Hierarchy
   - Autonomy of staff
   - Specialization
   - Top down versus two way communication
   - Vertical versus horizontal communication
   - Size

5. Management factors

   - Clarity of specification of goals, targets, and outputs
   - Clarity of specification of implementation plans, schedules, and responsibilities
   - Degree performance is rewarded and sanctioned
   - Information gathering and feedback
   - Multispeciality problem solving teams
   - Authoritarian or democratic management style
   - Flexible versus blue print programming

6. Process factors

    - Creating consensus on goals
    - Coordinating workers
    - Coordinating inter organizational relationships
    - Beneficiary participation
    - Experimentation
    - Analysis (reflection)
    - Planning
    - Adaptation to environment
    - Production

7. Strategy factors (choice between)

    - Performance and reflection
    - Quality and quantity outputs
    - Multiple and single outputs
    - Internal and external attention
    - Flexible and blueprint designs
    - Conservative and risky goals
    - High and low intensive marketing

8. Environment factors

    - National commitment
    - Stakeholder support and opposition
    - Stability (political, economic, social)
    - Conduciveness and hostility (political, economic, and social)
    - Permissiveness
    - Price distortion
    - National economic health
    - Director's and members political connections and networks

Source: Kurt Finsterbusch, University of Maryland. *Studying Success Factors in Multiple Cases Using Low-Cost Methods*, unpublished paper, 1990.

Gender Analysis

Definition

Gender analysis focuses on understanding and documenting the differences in gender roles, activities, and needs, in a given context. Gender analysis involves the disaggregation of qualitative and quantitative data by gender. It highlights the special roles and learned behavior of men and women based on gender attributes rather than biological attributes. These vary across cultures, class, ethnicity, income, education, and time. Thus gender analysis does not treat women as a homogeneous group nor gender attributes as permanent.

Intra-household dynamics

In the past, the household has often been the smallest unit of analysis. The assumption was, what is good for the household is good for all its members. However, the household itself is a system of resource allocation. All members of a household—men, women, and children—have different roles, skills, interests, needs, priorities, access, and control over resources. Thus, any development intervention which affects one member of the household will positively or negatively affect all others. Hence the importance of understanding these interdependent relationships, the rights, responsibilities, obligations, and patterns of interaction among household members. These interactions change with stages of the life cycle, shifts in external economic incentives, and vary from culture to culture.

Inter-household relations

Individuals and households belong to larger corporate groupings (clan, groups, temples, communities). Thus, individuals are members of larger groups with whom they are involved in labor exchanges, flow of goods, and other alliances for survival. It is important to understand the social organization of these larger networks and the gender differences in roles, functions, and access.

Purpose

Applied to development interventions, gender analysis helps:

- Identify gender-based differences in resource access in order to predict how different members of households, groups, and societies will participate in, and be affected by, planned development interventions.

- Allow planners to design policy reform and supportive program strategies which are effective, efficient, equitable, and empowering.
- Develop training packages to sensitize development staff on gender issues and develop training strategies for beneficiaries.

The gender analysis framework

Gender analysis is holistic (gender relations within social, organizational, economic and political life), relational (relationship between men and women) and complex (gender interacts with other factors such as age, class, ethnicity and income).

Many variants of policy and sector-specific gender analysis tools are available. However, the last decade of experience establishes that availability of gender-specific checklists and collecting gender-specific data by itself may not change anything unless institutional and planning approaches change and gender is integrated into the mainstream planning and design mode.

Gender planning is a mainstream planning activity and not a separate planning process focusing on women. Hence, the composition of the planning team, timing of data collection, the tabling of issues and integration of gender concerns into overall objectives is critical early in policy and project formulation. There are five key principles that need to be followed to ensure that gender analysis contributes to gender responsive policy and design change. They are:

Planning as a process. Programs that aim to be gender responsive must adapt flexible, interactive planning processes that are iterative, adjust objectives based on feedback and enable beneficiaries to be active participants in the planning process.

Gender diagnosis. Data collected through gender analysis should be organized to highlight key gender problems, underlying causes of problems for men and women and the relationship between them.

Gender objectives. There is a need to clarify what gender problems will be addressed and what are the goals, both practical and strategic. It is important to negotiate consensus on objectives at policy, managerial and working levels.

Gender strategy. Operational strategies which will be used to achieve stated objectives must be developed, along with appropriate incentives, budget, staff, training and organizational strategies.

Gender monitoring and evaluation. Flexible planning requires gender monitoring and evaluation to enable adjustment to experience and to establish accountability of commitment to achieve gender-specific priorities.

Categories of gender analysis

Five major categories of information comprise gender analysis. The extent to which information is collected depends on the nature of the problems being addressed and the quality and depth of information already available. The five categories of profiles are:

1. Needs assessment:

   - What are the priority needs of men and women
   - How can these needs be addressed
   - Which need can be solved at the local level, which require external intervention

2. Activities profile:

   - Who does what
   - What do women, men and children do
   - Time allocation, when do they do it, daily and seasonal
   - Where is the activity performed, location
   - How flexible or rigid is the division of labor

3. Resources access and control profile:

   - What resources are available to men and women to conduct their activities
   - What resources (land, produce, knowledge, cash, tools, institutions) do women and men have access to (and use)
   - Which resources do they control, the power to decide whether a resource is used, how it is used and how it is allocated

4. Benefits and incentives analysis:

Benefit analysis refers to (3) above, and goes further to analyze who controls outputs or benefits. Incentive analysis taps into user preferences, values placed on output, and the risks involved, which affect motivation to do or not to do. The incentives include taste, risk, convenience, time savings, reduced conflict, marketability, prestige, byproducts, etc.

5. Institutional constraints and opportunities:
   - Policies, laws, regulations, procedures from the national agency level to the community level
   - Agency organization, responsiveness, activities, training, gender, mobility, and skills of personnel
   - Planning and evaluation procedures, information flow, and outreach strategies

Sources:
Caroline Moser. 1993. *Gender Planning and Development: Theory, Practice and Training.* Rutledge, London.
Deepa Narayan. 1994. *Gender Analysis Note.* ENVSP.
Wendy Wakeman. 1994. *Gender Issues Sourcebook for Water and Sanitation Projects.* World Bank. UNDP-World Bank Water and Sanitation Program.

ANNEXES

Annex 1

Terms of Reference

Mauritania: Agriculture, Forestry, and Livestock
Resources Management Project

Process of identification of a rural development
project by its future beneficiaries
(Participatory Project Preparation)

A. Background

1. The Sahelian Department of the World Bank has undertaken, in concert with several regional governments of the region, a new generation of rural development projects in which participation of beneficiaries is central. However, because of limited experience is use of participatory methods in project identification, a consultative process will be used, facilitated by local and expatriate consultants. The objective of these terms of reference is to permit their participation in the identification of a future project of which they will be the beneficiaries.

B. Objectives

2. The objective of the consultation is to give the personnel charged with the identification, preparation and evaluation of the project, (i.e., personnel of the FAO-CP, the World Bank, and the Mauritanian Administration), the information which will permit them to a) ensure the usefulness of the project in the proposed sector and b) to adjust its objectives, strategy, and the component parts to the expressed needs and the capacity of the partners, (i.e, the concerned communities and the government personnel charged with the execution of the project). It is because the consultation with the beneficiaries must be conducted concurrent with related technical services, that it is the responsibility of the consultant to keep a watch out for the links between the present consultation and the program of "Local Working Groups" established by the FAO-CP

3. The second element of the consultation is its connection to the environmental impact (EA) evaluation process upon which this type of project depends. This process takes place outside the project cycle, at the time that the zone of intervention is determined and the project components are sufficiently clear so that their impact, particularly on populations with affected zone, can be evaluated. Consequently, the

consultant will study in detail Operational Directives of the World Bank so as to assure that the results of the consultation can be utilized outside the course of the EA, in the form of a complimentary survey using a similar beneficiary sample.

4. The third element which the consultant should consider is the level of participation which can be expected on the part of the beneficiaries; considering the likelihood that the consultation leads to a project, and the degree of interest on the part of the beneficiaries, the consultant will create a training module focusing on the two elementary levels of participation: a) consulting with beneficiaries to obtain information, while giving them the possibility to express their point of view; and b) the choice of the beneficiaries among several scenarios and possible alternatives. The implementation of more advanced levels of participation such as a) autonomous decision-making; b) total responsibility in the development of resources, particularly financial, will be delayed until later stages of the project cycle.

C. Continuation of consultation

5. The implementation of the identification process by future beneficiaries is comprised of six different steps corresponding to the respective contributions of the national consultant and the expatriate specialist:

a) *Sampling of beneficiaries*; in coordination with Livestock Cooperative Association officials and taking into account the population now organized in these associations, the diversity of natural and social conditions, and the population which will be involved in the first phase of the project, the two consultants will indicate the of communities to be consulted, their distribution, the specific community members to be consulted, etc.

b) *Areas for further elaboration*; the expatriate consultant will have an understanding with the project identification and preparation team, regarding the nature of the information which will be elaborated upon as a result of the consultation with beneficiaries:

phase one:

- Community values, quality of life;
- Objectives sought, production activities;
- Description of local environment
- Evaluation of resources;
- How to realize objectives with available resources;
- Obstacles encountered;
- Priority problems to be resolved;

phase two:

- Existence, characteristics and function of social organizations;
- Level and areas of experience and expertise available within the community;
- Socio-cultural feasibility and acceptability of the solutions envisioned by the community;

phase three:

- Objectives of the future project;
- Social, economic and technical strategies;
- Project components;
- Estimated cost, eventual contribution of the community.

c) *Setting the agenda for the training sessions:* taking into account his/her own experience and the information existing in the literature, the expatriate consultant:

- Will establish the process by which the training sessions organized with the beneficiaries should permit the generation of previously identified information:

  - Session participants;
  - Duration of sessions;
  - Content: sequence of implementation exercises;

- Will select the exercises of the accelerated participatory research methods, or PRA in English, subject to the participation of the community, for example:

  - Group interviews;
  - Individual interviews;
  - Map and model making of the environment;
  - Classification of priorities according to target groups;
  - Games and simulations, etc.

d) *Testing and adjusting of the agenda:* the proposed initial agenda will be tested concurrently by the expatriate consultant and the national consultant in the actual conditions of the communities selected in the sample; the necessary adjustment will be provided;

e) *Program execution of the training sessions:* should be conducted by the national consultant who speaks the language of the specific community;

f) *Analysis, synthesis and presentation of results:* all of the information collected during the training sessions is developed (if possible in disk/software format) by the two consultants so that it can be utilized by the personnel charged with the design and preparation of the project.

D. Products of the consultation

6. The following documents will be produced by the consultation team:

- First of all, each consultant will write a short report of 4 to 5 pages in which he/she will present his/her own perception of the consultation and how he/she should proceed to obtain the best possible results.
- Sampling of the communities to be consulted;
- List of themes to establish;
- Tentative agenda, and the final agenda of the community training sessions;
- If a questionnaire is deemed necessary, it should be as short as possible;
- Calendar of the implementation of training sessions;
- If possible, information should be available on disk/computer;
- Synthesis of the information established in the course of the consultation, ... the final document

E. Profile of consultants

7. *Expatriate consultant:* with a specialty in the group training, particularly in the rural context, with, if possible, experience in North Africa and/or the southern Sahara for which he has created modules directed toward different target groups and with various objectives; the consultant is also a specialist in the involvement of adults and an expert in PRA methods.

8. *National consultant:* an experienced trainer, familiar with the communities in which the training sessions will take place, fluent in the language of the community; he/she should also have a good level of experience in PRA techniques.

F. Duration of the mission and tentative agenda

9. Expatriate consultant:

Sampling, design and testing of methodological module: 1 week in country of origin, 1 week in Mauritania, in April 1994. Return ticket from country of origin to Mauritania.

Synthesis, analysis and presentation of information: 2 weeks in country of origin, 1 week in Mauritania, in July 1994; Return ticket from country of origin to Mauritania.

10. National consultant:

Assist in the sampling, design and testing of module: 1 week April 1994;

Implementation of training sessions: 3 weeks between May and June 1994.

Assist in the synthesis, analysis and presentation of results: 1 week in Mauritania.

G. Documents to be sent to consultants

- Description of project;

- Memo on beneficiary participation in the natural resources management projects;

- World Bank's Operational Directive on environmental impact evaluations;

- Terms of reference for the Local Working Groups.

Source: "Mauritanie: Projet de Gestion des Ressources Agro-sylvo-pastorale." Termes de Reference. La Banque Mondiale.

Annex 2

Bibliography and Selected References

Baron, Mark, Zollinger, Marcel & Brocklehurst, Clarissa. 1989. *Rapport D'Évaluation: CARE Cameroun Project D'Eau Potable et D'Éducation sur la Santé Communautaire Province de L'Est.* Préparé pour CARE Canada par Cowater International Inc., Ottawa, Canada.

Boyle, N. and Wight, A. 1992. "Policy Reform: The Role of Informal Organizations." World Bank. Infrastructure and Urban Development Department. Urban No. OU-5, September.

Chambers, Robert. 1993. "A Note for World Bank Staff on Participatory Rural Appraisal."

Chambers, Robert. 1993. "Challenging the Professions: Frontiers for Rural Development." ITDG, London.

Curtis, Valerie. 1988. "Evaluation of UNICEF Assisted Water Programmes in the People's Republic of China." London School of Hygiene and Education.

Debebe, Dessalegn. 1990. "The Role of Community Participants in RRA Methods in Ethiopia." In: *RRA Notes*, No. 8 January 1990. International Institute for Environment and Development: London, UK.

Evans, P.A. 1985. *Monitoring and Evaluation of the Lesotho National Rural Sanitation Programme Draft Plan of Operations.* Ministry of Health: Lesotho.

Final Review of Case Studies of Women's Participation in Community Water Supply and Sanitation. Report on Workshop, WHO and PROWWESS. Kupang Indonesia, 23-27 May, 1988.

Finsterbusch, Kurt. 1990. "Studying Success Factors in Multiple Cases Using Low Cost Methods." Unpublished. Presented at the XII World Congress of Sociology, Madrid, July 1990.

International Council for Adult Education. 1981. *Drawing from Action for Action: Drawing and Discussion as a Popular Research Tool.* Working Paper/Participation Research Project; 6: Toronto, Ontario.

IRC International Water and Sanitation Centre (1988). *Community Participation and Women's Involvement in Water Supply and Sanitation Projects.* Occasional Paper Series, no. 12. IRC/WHO Collaborating Centre: The Hague, Netherlands.

Kapoor, Kapil, *et al.* 1993. *Uganda: Growing Out of Poverty.* The World Bank: Washington, DC.

Malone Given Parsons Ltd. 1985. *Ghana Upper Region Water Programme Evaluation Project. Report 5: Technical Appendix One, Survey Methodology.* Prepared for Canadian International Development Agency.

Moser, Caroline. 1993. Gender Planning and Development: Theory, Practice and Training. Routledge, London.

Narayan, D. and Srinivasan, L. 1994. *Participatory Development Tool Kit.* The World Bank: Washington, DC.

Narayan, Deepa. 1993. *Participatory Evaluation: Tools for Managing Change in Water and Sanitation.* World Bank Technical Paper number 207. The World Bank: Washington, DC.

Narayan, Deepa. 1994. "Participatory Gender Analysis Tools for the Community and Agency Levels." Unpublished paper. ENVSP. The World Bank: Washington, DC.

Narayan, Deepa. 1994. "The Contribution of People's Participation: Evidence from 121 Rural Water Supply Projects." ESD, occasional paper Series No.1. World Bank.

Nyagah, Justin Mugo. 1992. "Paper on the Experience of PLAN/Embu on the Application of SARAR Technique During the Final Borehole Evaluation."

Ole Shani, Daniel and Perkins, Mitali 1991. "Participation in Development: Learning from the Maasai People's Program in Kenya. World Vision Staff Working Paper No. 12. World Vision International: Monrovia, CA.

Rajathi, C. 1992. IRSENS Project Danida, Tamil Nadu Report. Adapted by ACDIL from Report presented at the Danida/World Bank/ACDIL workshop in Goa, India, 1992.

Schaefer Davis, S., Fatine, N., Alaoui, C. 1993. "Étude socio-économique du projet d'AEP des Zone Rurales de la Vallée du Ziz et de la Plaine du Tafilalet." Vol. 1. IBRD, UNDP, Office National de l'Eau Potable, Morocco.

Seidl, Peter and B.J. Dutka. 1993. Developing a Self-Sustained Microbiological Water Quality Testing Capability in a Remote Aboriginal Community.

Srinivasan, L., Roshaneh, Z., & Minnatullah, K.M. 1994. *Community Participation: Strategies and Tools. A Trainers' Manual for the Rural Water Supply and Sanitation Sector in Pakistan.* Ministry of Local Government and Rural Development, Government of Pakistan;

United Nations Development Programme; United Nations Children's Fund; UNDP/World Bank RWSG-SA.

Srinivasan, Lyra . 1992. *Options for Educators: A Monograph for Decision Makers on Alternative Participatory Strategies.* PACT/CDS: New York, NY.

Srinivasan, Lyra. 1990. *Tools for Community Participation: A Manual for Training Trainers in Participatory Techniques.* PROWWESS/UNDP: New York, NY.

Wakeman, Wendy. 1993. Gender Issues Sourcebook for Water and Sanitation. World Bank: Washington, DC.

Walker, Horatio. 1991. "Who Interprets? Who Decides? Participatory Evaluation in Chile." in: *Development Communications Report*, no. 72, 1991/1.

Welbourn, Alice. "RRA and the Analysis of Difference" in *RRA Notes: Participatory Methods for Learning and Analysis*, No. 14, (pp. 14-23). December 1991. International Institute for Environment and Development: London, UK.

Whiteford, Linda. 1993. "Women's Voices Heard in Ecuador Health Risk Assessment." in *Voices from the City: Newsletter of the Peri-Urban Network on Water supply and Environmental Health Sanitation*, vol 3, September 1993

World Bank 1993. "Karnataka Rural Water Supply and Environmental Sanitation Project." *Staff Appraisal Report, India.* March 31, 1993. Infrastructure Operations Division. Country Department II - India. South Asia Regional Office. The World Bank: Washington, DC.

Young, Helen. 1990. Use of Wealth Ranking in Nutrition Surveys in Sudan. in *RRA Notes*, No. 8. International Institute for Environment and Development: London, UK.

World Bank Publications

Prices and credit terms vary from country to country. Consult your local distributor before placing an order.

ALBANIA
Adrion Ltd.
Perlat Rexhepi Str.
Pall. 9, Shk. 1, Ap. 4
Tirana
Tel: (42) 274 19; 221 72
Fax: (42) 274 19

ARGENTINA
Oficina del Libro Internacional
Av. Cordoba 1877
1120 Buenos Aires
Tel: (1) 815-8156
Fax: (1) 815-8354

AUSTRALIA, FIJI, PAPUA NEW GUINEA, SOLOMON ISLANDS, VANUATU, AND WESTERN SAMOA
D.A. Information Services
648 Whitehorse Road
Mitcham 3132
Victoria
Tel: (61) 3 9210 7777
Fax: (61) 3 9210 7788
URL: http://www.dadirect.com.au

AUSTRIA
Gerold and Co.
Graben 31
A-1011 Wien
Tel: (1) 533-50-14-0
Fax: (1) 512-47-31-29

BANGLADESH
Micro Industries Development
Assistance Society (MIDAS)
House 5, Road 16
Dhanmondi R/Area
Dhaka 1209
Tel: (2) 326427
Fax: (2) 811188

BELGIUM
Jean De Lannoy
Av. du Roi 202
1060 Brussels
Tel: (2) 538-5169
Fax: (2) 538-0841

BRAZIL
Publicações Tecnicas Internacionais Ltda.
Rua Peixoto Gomide, 209
01409 Sao Paulo, SP.
Tel: (11) 259-6644
Fax: (11) 258-6990

CANADA
Renouf Publishing Co. Ltd.
1294 Algoma Road
Ottawa, Ontario K1B 3W8
Tel: 613-741-4333
Fax: 613-741-5439

CHINA
China Financial & Economic Publishing House
8, Da Fo Si Dong Jie
Beijing
Tel: (1) 333-8257
Fax: (1) 401-7365

COLOMBIA
Infoenlace Ltda.
Apartado Aereo 34270
Bogotá D.E.
Tel: (1) 285-2798
Fax: (1) 285-2798

COTE D'IVOIRE
Centre d'Edition et de Diffusion
Africaines (CEDA)
04 B.P. 541
Abidjan 04 Plateau
Tel: 225-24-6510
Fax: 225-25-0567

CYPRUS
Center of Applied Research
Cyprus College
6, Diogenes Street, Engomi
P.O. Box 2006
Nicosia
Tel: 244-1730
Fax: 246-2051

CZECH REPUBLIC
National Information Center
prodejna, Konviktska 5
CS – 113 57 Prague 1
Tel: (2) 2422-9433
Fax: (2) 2422-1484
URL: http://www.nis.cz/

DENMARK
SamfundsLitteratur
Rosenoerns Allé 11
DK-1970 Frederiksberg C
Tel: (31)-351942
Fax: (31)-357822

ECUADOR
Facultad Latinoamericana de
Ciencias Sociales
FLASCO-SEDE Ecuador
Calle Ulpiano Paez 118
y Av. Patria
Quito, Ecuador
Tel: (2) 542 714; 542 716; 528 200
Fax: (2) 566 139

EGYPT, ARAB REPUBLIC OF
Al Ahram
Al Galaa Street
Cairo
Tel: (2) 578-6083
Fax: (2) 578-6833

The Middle East Observer
41, Sherif Street
Cairo
Tel: (2) 393-9732
Fax: (2) 393-9732

FINLAND
Akateeminen Kirjakauppa
P.O. Box 23
FIN-00371 Helsinki
Tel: (0) 12141
Fax: (0) 121-4441
URL: http://booknet.cultnet.fi/aka/

World Bank Publications
66, avenue d'Iéna
75116 Paris
Tel: (1) 40-69-30-56/57
Fax: (1) 40-69-30-68

GERMANY
UNO-Verlag
Poppelsdorfer Allee 55
53115 Bonn
Tel: (228) 212940
Fax: (228) 217492

GREECE
Papasotiriou S.A.
35, Stournara Str.
106 82 Athens
Tel: (1) 364-1826
Fax: (1) 364-8254

HONG KONG, MACAO
Asia 2000 Ltd.
Sales & Circulation Department
Seabird House, unit 1101-02
22-28 Wyndham Street, Central
Hong Kong
Tel: 852 2530-1409
Fax: 852 2526-1107
URL: http://www.sales@asia2000.com.hk

HUNGARY
Foundation for Market Economy
Dombovari Ut 17-19
H-1117 Budapest
Tel: 36 1 204 2951 or
36 1 204 2948
Fax: 36 1 204 2953

INDIA
Allied Publishers Ltd.
751 Mount Road
Madras - 600 002
Tel: (44) 852-3938
Fax: (44) 852-0649

INDONESIA
Pt. Indira Limited
Jalan Borobudur 20
P.O. Box 181
Jakarta 10320
Tel: (21) 390-4290
Fax: (21) 421-4289

IRAN
Kowkab Publishers
P.O. Box 19575-511
Tehran
Tel: (21) 258-3723
Fax: 98 (21) 258-3723

Ketab Sara Co. Publishers
Khaled Eslamboli Ave.,
6th Street
Kusheh Delafrooz No. 8
Tehran
Tel: 8717819 or 8716104
Fax: 8862479

IRELAND
Government Supplies Agency
Oifig an tSoláthair
4-5 Harcourt Road
Dublin 2
Tel: (1) 461-3111
Fax: (1) 475-2670

ISRAEL
Yozmot Literature Ltd.
P.O. Box 56055
Tel Aviv 61560
Tel: (3) 5285-397
Fax: (3) 5285-397

R.O.Y. International
PO Box 13056
Tel Aviv 61130
Tel: (3) 5461423
Fax: (3) 5461442

Palestinian Authority/Middle East
Index Information Services
P.O.B. 19502 Jerusalem
Tel: (2) 271219

ITALY
Licosa Commissionaria Sansoni SPA
Via Duca Di Calabria, 1/1
Casella Postale 552
50125 Firenze
Tel: (55) 645-415
Fax: (55) 641-257

JAMAICA
Ian Randle Publishers Ltd.
206 Old Hope Road
Kingston 6
Tel: 809-927-2085
Fax: 809-977-0243

JAPAN
Eastern Book Service
Hongo 3-Chome,
Bunkyo-ku 113
Tokyo
Tel: (03) 3818-0861
Fax: (03) 3818-0864
URL: http://www.bekkoame.or.jp/~svt-ebs

KENYA
Africa Book Service (E.A.) Ltd.
Quaran House, Mfangano Street
P.O. Box 45245
Nairobi
Tel: (2) 23641
Fax: (2) 330272

KOREA, REPUBLIC OF
Daejon Trading Co. Ltd.
P.O. Box 34
Yeoeida
Seoul
Tel: (2) 785-1631/4
Fax: (2) 784-0315

MALAYSIA
University of Malaya Cooperative
Bookshop, Limited
P.O. Box 1127
Jalan Pantai Baru
59700 Kuala Lumpur
Tel: (3) 756-5000
Fax: (3) 755-4424

MEXICO
INFOTEC
Apartado Postal 22-860
14060 Tlalpan,
Mexico D.F.
Tel: (5) 606-0011
Fax: (5) 606-0386

NETHERLANDS
De Lindeboom/InOr-Publikaties
P.O. Box 202
7480 AE Haaksbergen
Tel: (53) 574-0004
Fax: (53) 572-9296

New Market
Auckland
Tel: (9) 524-8119
Fax: (9) 524-8067

NIGERIA
University Press Limited
Three Crowns Building Jericho
Private Mail Bag 5095
Ibadan
Tel: (22) 41-1356
Fax: (22) 41-2056

NORWAY
Narvesen Information Center
Book Department
P.O. Box 6125 Etterstad
N-0602 Oslo 6
Tel: (22) 57-3300
Fax: (22) 68-1901

PAKISTAN
Mirza Book Agency
65, Shahrah-e-Quaid-e-Azam
P.O. Box No. 729
Lahore 54000
Tel: (42) 7353601
Fax: (42) 7585283

Oxford University Press
5 Bangalore Town
Sharae Faisal
PO Box 13033
Karachi-75350
Tel: (21) 446307
Fax: (21) 454-7640

PERU
Editorial Desarrollo SA
Apartado 3824
Lima 1
Tel: (14) 285380
Fax: (14) 286628

PHILIPPINES
International Booksource Center Inc.
Suite 720, Cityland 10
Condominium Tower 2
H.V dela Costa, corner
Valero St.
Makati, Metro Manila
Tel: (2) 817-9676
Fax: (2) 817-1741

POLAND
International Publishing Service
Ul. Piekna 31/37
00-577 Warzawa
Tel: (2) 628-6089
Fax: (2) 621-7255

PORTUGAL
Livraria Portugal
Rua Do Carmo 70-74
1200 Lisbon
Tel: (1) 347-4982
Fax: (1) 347-0264

ROMANIA
Compani De Librarii Bucuresti S.A.
Str. Lipscani no. 26, sector 3
Bucharest
Tel: (1) 613 9645
Fax: (1) 312 4000

RUSSIAN FEDERATION
Isdatelstvo <Ves Mir>
9a, Kolpachniy Pereulok
Moscow 101831
Tel: (95) 917 87 49
Fax: (95) 917 92 59

SAUDI ARABIA, QATAR
Jarir Book Store
P.O. Box 3196
Riyadh 11471
Tel: (1) 477-3140
Fax: (1) 477-2940

SINGAPORE, TAIWAN, MYANMAR, BRUNEI
Ashgate Publishing Asia
Pacific Pte. Ltd.
41 Kallang Pudding Road #04-03
Golden Wheel Building
Singapore 349316
Tel: (65) 741-5166
Fax: (65) 742-9356
e-mail: ashgate@asianconnect.com

SLOVAK REPUBLIC
Slovart G.T.G. Ltd.
Krupinska 4
PO Box 152
852 99 Bratislava 5
Tel: (7) 839472
Fax: (7) 839485

SOUTH AFRICA, BOTSWANA
For single titles:
Oxford University Press
Southern Africa
P.O. Box 1141
Cape Town 8000
Tel: (21) 45-7266
Fax: (21) 45-7265

For subscription orders:
International Subscription Service
P.O. Box 41095
Craighall
Johannesburg 2024
Tel: (11) 880-1448
Fax: (11) 880-6248

SPAIN
Mundi-Prensa Libros, S.A.
Castello 37
28001 Madrid
Tel: (1) 431-3399
Fax: (1) 575-3998
http://www.tsai.es/mprensa

Mundi-Prensa Barcelona
Consell de Cent, 391
08009 Barcelona
Tel: (3) 488-3009
Fax: (3) 487-7659

SRI LANKA, THE MALDIVES
Lake House Bookshop
P.O. Box 244
100, Sir Chittampalam A.
Gardiner Mawatha
Colombo 2
Tel: (1) 32105
Fax: (1) 432104

SWEDEN
Fritzes Customer Service
Regeringsgaton 12
S-106 47 Stockholm
Tel: (8) 690 90 90
Fax: (8) 21 47 77

P. O. Box 1305
S-171 25 Solna
Tel: (8) 705-97-50
Fax: (8) 27-00-71

SWITZERLAND
Librairie Payot
Service Institutionnel
Côtes-de-Montbenon 30
1002 Lausanne
Tel: (021)-320-2511
Fax: (021)-311-1393

Van Diermen Editions Techniq
Ch. de Lacuez 41
CH1807 Blonay
Tel: (021) 943 2673
Fax: (021) 943 3605

TANZANIA
Oxford University Press
Maktaba Street
PO Box 5299
Dar es Salaam
Tel: (51) 29209
Fax: (51) 46822

THAILAND
Central Books Distribution
306 Silom Road
Bangkok
Tel: (2) 235-5400
Fax: (2) 237-8321

TRINIDAD & TOBAGO, JAM.
Systematics Studies Unit
#9 Watts Street
Curepe
Trinidad, West Indies
Tel: 809-662-5654
Fax: 809-662-5654

UGANDA
Gustro Ltd.
Madhvani Building
PO Box 9997
Plot 16/4 Jinja Rd.
Kampala
Tel/Fax: (41) 254763

UNITED KINGDOM
Microinfo Ltd.
P.O. Box 3
Alton, Hampshire GU34 2PG
England
Tel: (1420) 86848
Fax: (1420) 89889

ZAMBIA
University Bookshop
Great East Road Campus
P.O. Box 32379
Lusaka
Tel: (1) 213221 Ext. 482

ZIMBABWE
Longman Zimbabwe (Pte.)Ltd
Tourle Road, Ardbennie
P.O. Box ST125
Southerton
Harare
Tel: (4) 662711
Fax: (4) 662716